Fabric Manufacturing Technology

Fabric Manufacturing Technology
Weaving and Knitting

K. Thangamani and S. Sundaresan

CRC Press
Taylor & Francis Group
Boca Raton London

CRC Press is an imprint of the
Taylor & Francis Group, an **informa** business

First edition published 2022
by CRC Press
6000 Broken Sound Parkway NW, Suite 300, Boca Raton, FL 33487–2742

and by CRC Press
4 Park Square, Milton Park, Abingdon, Oxon, OX14 4RN

© 2022 Taylor & Francis Group, LLC

CRC Press is an imprint of Taylor & Francis Group, LLC

Library of Congress Cataloging-in-Publication Data
A catalog record for this book has been requested

ISBN: 978-0-367-42583-8 (hbk)
ISBN: 978-0-367-42585-2 (pbk)
ISBN: 978-0-367-85368-6 (ebk)

DOI: 10.1201/9780367853686

Typeset in Times
by Apex CoVantage, LLC

This work is dedicated to

*my late father R. Kandasamy, mother Chinnammal,
Uncle A. Subramaniam*

and my brothers and sisters

who have given me inspiration and support

throughout my educational, and professional life.

Dr. K. Thangamani

Contents

Knitting

Nonwoven

Preface

Even though fabric manufacturing techniques are as old as human civilization, they took different forms with different raw materials as civilization grew. In prehistoric days, humans made clothing using leaves. As civilization grew, the raw materials, as well as the techniques, changed. About 5,000 years ago, the material for clothing changed from leaves to cotton and silk. The process of weaving by hand was developed. The weft yarn was inserted using sticks between the top and bottom layers of warp yarn by hand. It was an art rather than a science. Human skill played a major role in producing different quality fabrics.

In 1733, Kay made a revolution by inventing the fly shuttle. The art aspect decreased, and the science aspect increased. In 1785, Cartwright transformed fabric manufacturing into a science and technology by inventing the power loom. For about two centuries, the basics of weaving were the same until shuttleless weaving was invented in the last quarter of the 20th century. Similarly, even though knitting was known five centuries ago, only in the last 50 years has it emerged as a method of fabric production. Other than weaving and knitting, the nonwoven method also has emerged in a big way.

The earlier books written on weaving and weaving preparatory processes were very elaborate on the details of the mechanism of weaving machines. The advent of electronics and developments in materials changed the very nature and principles of mechanism. Hence, in today's weaving machines, the mechanisms are totally different in nature. Therefore, the earlier books became irrelevant. A necessity has arisen for a new book on fabric manufacture that is concise and at the same time explains the basic principles precisely. This book, *Fabric Manufacturing Technology*, is an attempt to meet this necessity.

All three major methods of fabric manufacture, namely, weaving, knitting and nonwoven have been dealt with in this book. Since conventional power looms and pirn-changing automatic looms are giving way to shuttleless looms, no attempt is made to describe them in detail. The shuttleless loom, with the necessary diagrams, is described in detail in the weaving section. The principles of weft insertion, the shedding mechanism, beat-up mechanisms for projectile, rapier, and air-jet looms are dealt with in the weaving section. Auxiliary motions like let-off and selvedge motion are also discussed. The basic concepts and principles of weft and warp knitting are discussed in the knitting section. The methods of batt preparation and bonding techniques are briefly dealt with in the nonwoven section.

The authors, with two decades of teaching and a decade of industrial experience, started writing the book with the thought that this book should be useful to the students of textile technology and fashion technology in undergraduate and polytechnic colleges. It will also serve as a reference book for understanding the fabric manufacturing process for professionals who are working in the garment industry.

Dr. K. Thangamani

Authors

Dr. K. Thangamani completed his B. Tech degree in 1981 at PSG College of Technology, Coimbatore, India and started his career as a trainee engineer at Textool Company Ltd, Coimbatore. Beginning in 1981, he worked for four years at the National Textile Corporation Ltd., Bangalore, first as a management trainee and then as an assistant technical manager. In 1985, he joined Tamarai Mills Ltd Coimbatore as an assistant spinning master. From 1988–1990, he was with PSG Polytechnic in research projects. He completed his M. Tech from Bharathiar University in 1993. Between 1990 and 1997, he joined PSG College of Technology as a project officer with the United Nations Development Programme and Government of India—assisting the National Jute Development Project and conducting more than fifty training programs and ten seminars/conferences throughout south India for promotion of diversified jute products. The jute shopping bags were introduced at that time. He was with NIFT TEA Knitwear Fashion Institute, Tirupur as an assistant director for a brief period.

In 1998, he joined Kumaraguru College of Technology, Coimbatore as a faculty member in the Department of Textile Technology and retired as professor of textile technology in 2020. His areas of interest are knitting technology, nonwoven, composites, and fabric manufacture. He obtained his PhD in textile technology in 2007 from Bharathiar University and is a recognized supervisor for PhD scholars at Anna University, Chennai. Under his supervision, five scholars have obtained their PhDs from Anna University, Chennai and one scholar from Bharathiar University, Coimbatore.

He was executive committee member of the Institution of Engineers (India), Tamil Nadu State Center for eight years from 2002–2010. He was organizing secretary for the International Conference on "Advancements in Specialty Textiles and Their Applications in Material Engineering and Medical Sciences" organized jointly with Technical University of Liberec, Czech Republic at Kumaraguru College of Technology, Coimbatore in 2014. He was the joint organizing secretary for the International Conference on "Waste Water Management" organized by the Departments of Textile Technology, Bio Technology and Civil Engineering with Tel Aviv University, Israel and Technical University of Liberec, Czech Republic, at Kumaraguru College of Technology, Coimbatore in 2017.

He is one of the editors for *Waste Water Management*, which contains selected articles from the International Conference on "Waste Water Management" conducted in 2017 at the Kumaraguru College of Technology. He published more than 25 research articles in International journals. He is a reviewer for three International research journals and reviewed more than 75 manuscripts.

Dr. S. Sundaresan has completed his B. Tech, M. Tech and PhD in textile technology and MBA in business administration. He has worked in textile industry as a supervisor for 10 years. He has more than 15 years of teaching experience, and at present, is working as associate professor in the Department of Textile Technology, Kumaraguru College of Technology, Coimbatore. He has published more than 20 papers in leading journals. He has co-authored *Home Furnishing* (Woodhead Publishing, India) and has undertaken consultancy works in the industry.

1 Introduction

1.1 CLOTHING

Clothing is one of the three basic needs of human beings, the other two needs being food and shelter. The production of clothing was carried out in one form or another from the early days of civilization. In olden days, the materials used for clothing production were mainly natural fibres, such as cotton, wool and silk. While cotton is cultivated from the cotton plant, wool is taken from sheep and silk from the silkworm. The cotton and wool fibres themselves cannot be used directly to form fabrics. These have to be converted into an intermediate product, namely yarn, and the yarn is converted into fabrics. The process of converting the yarn into fabric is called weaving, and the equipment used for weaving is called a loom. In earlier days, other than weaving, fabric was also made from yarn by hand knitting using two pins.

Even though clothing is the main application for textile fabrics, nowadays, the uses of textile fabrics are found in every field. Specially manufactured textile fabrics are extensively used in hygiene products in the medical field, environmental protection, transportation, geotextiles, conveyor belts and safety and protective fabrics, among others.

1.2 RAW MATERIALS

Even though natural fibres were available abundantly, synthetic raw materials were developed to augment the properties of natural raw materials for fabric production from the middle of 20th century. Of the synthetic raw materials polyester, nylon and rayon are used extensively for fabric production. The raw materials for textile applications should be in the form of fibres. The fibres will have very large length to width/thickness ratio; only then can they be used for textile application. The textile fibres should have strength, elongation, flexibility, abrasion resistance and moisture absorption for better performance in fabrics.

The fibres used for textile applications are:
Natural fibres: Cotton, linen, jute, hemp, ramie, wool and silk
Synthetic fibres: Polyester, nylon, rayon, spandex, olefin, aramid, acrylic, carbon and glass

1.3 SPINNING

The fibre length of cotton varies from 12mm to 38mm, and the fibre length of wool varies from 50mm to 100mm. The synthetic fibres can be manufactured in required lengths or as filaments form. The process of converting the fibres into yarn is called spinning. Silk is available naturally in filament form, and it does not require spinning. There are two types of spinning mainly used for producing cotton yarn: (1) ring spinning and (2) open-end spinning.

1.3.1   RING SPINNING

Spinning is a process in which the fibre strands are aligned parallel and twisted. The twisting of the strand of fibres increases the friction forces between the fibres and gives strength to the strand to make it a yarn. Twisting the strand is done by a ring and traveller system and this is called ring spinning. Figure 1.1 shows a ring spinning system.

1.3.2   OPEN-END SPINNING

In open-end spinning, the fibres are fed into the opening roller in the form of a sliver. The opening roller opens the fibres and feeds the fibres in the form of a strand to the rotor which is rotating at very high speed. The rotating rotor gives twist to the strand and the yarn is taken out. Because a rotor is used to give the twist, this is also called rotor spinning. Rotor spinning is suitable to produce course yarns from short fibres. Figure 1.2 shows a rotor spinning system.

The fibre arrangement in a stable fibre yarn is shown in Figure 1.3, and a photograph of yarn is shown in Figure 1.4

1.4   YARN NUMBERING SYSTEM

Yarn is made from short fibres and the uniformity of the yarn depends on the uniform placement of fibres in its axis. This is not achieved fully. Its diameter varies in its cross section. Moreover measuring the yarn diameter is difficult due to its structure. Therefore, a linear density measurement is used to number the yarn

There are two methods by which yarn is numbered

1. *Indirect system*

   This is the traditional system that originated from England. In this system length per unit weight is specified. The number of hanks (840 yards) present in 1 pound of yarn is defined as the yarn number or yarn count. Suppose in a particular yarn if 10 hanks (8,400 yards) weighs 1 pound, then the yarn count of that yarn will be 10s Ne. ("Ne" refers to English count.) In the indirect system, the higher the yarn number or count, the finer will be the yarn. For example, 40s Ne yarn will be finer than 20s Ne yarn.

2. *Direct system*

   In the direct system, the weight per unit length is specified. The weight in grams per 1000 metres of yarn is defined as the yarn count and this count is called tex. For example if 1000 metres of yarn weighs 40 grams, then

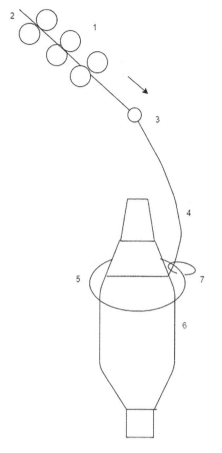

1. Drafting system
2. Fibre strand
3. Yarn guide
4. Yarn
5. Ring
6. Bobbin
7. Traveller

FIGURE 1.1   Ring spinning system.

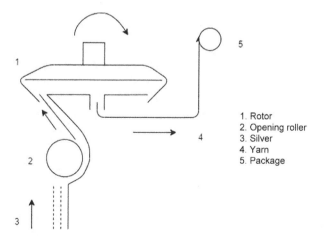

1. Rotor
2. Opening roller
3. Silver
4. Yarn
5. Package

FIGURE 1.2   Rotor spinning system.

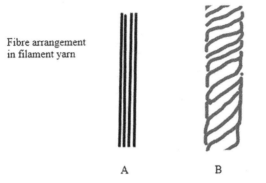

Fibre arrangement
in filament yarn

Fibre arrangement
in spun yarn

A   B

FIGURE 1.3   Fibre arrangement in yarn.

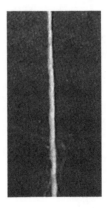

FIGURE 1.4   Photograph of yarn.

the yarn count is 40tex. In the direct system, the higher the yarn count number, the courser will be the yarn. For numbering filaments and fibres, another term is used: denier. *Denier* is defined as the weight in grams of 9000 metres of filament yarn.

2 Fabric Formation Methods

Fabrics are made by assembling yarn or fibres or a combination of both. They are assembled in such a way that due to their inter fibre frictional force, they produce a material that has strength, elongation, flexibility, abrasion resistance and other properties suitable for clothing and other end uses. There are mainly three methods by which fabrics can be produced for major applications. Figure 2.1 shows the fabric formation methods. Each method has its own merits and demerits, and each has its own end uses.

The three main fabric formation methods are

1. Weaving,
2. Knitting and
3. Nonwoven.

2.1 WEAVING

Weaving is a process in which two sets of yarns are interlaced to form a fabric. The lengthwise set of yarn is called warp, and the widthwise yarn is called weft. To make fabric, the lengthwise yarn and the widthwise yarn have to be interlaced in such a way that the warp yarns pass under and over the weft yarns in a systematic order. Similarly, the weft passes under and over the warp yarns in a systematic order. Figure 2.2 show the passage of warp and weft yarn in woven fabric. Figure 2.3 shows a woven fabric structure.

During weaving, the warp yarns are split into two layers which is called a shed. The weft yarn is inserted into the shed and pushed to the point where the warp becomes the cloth. In earlier days, the method employed for the insertion of weft was by means of a stick with a hooked end. The stick was inserted first in one direction and then in the other direction.

In the middle of 18th century, a fly shuttle with a weft package was invented to insert the weft inside the warp shed. The shuttle was thrown through the warp shed from one side and then from the other side. This eliminated the weft insertion by sticks and made a revolution in fabric making. Initially, propelling the shuttle from one side to the other was done manually using a wooden frame called a handloom. After the invention of steam power, the shuttle was propelled mechanically using steam power. This weaving frame was called a power loom. Since the middle of 20th century, attempts have been made to replace the shuttle for weft insertion, and as a result, nowadays shuttleless looms share bulk of fabric production.

DOI: 10.1201/9780367853686-2

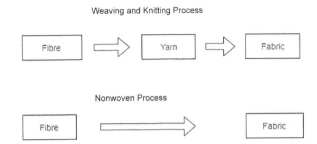

FIGURE 2.1   Fabric formation methods.

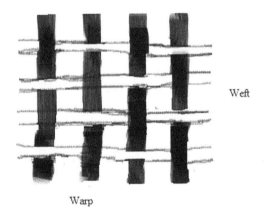

FIGURE 2.2   Passage of warp and weft yarn in woven fabric.

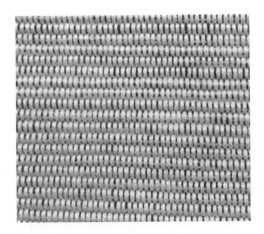

FIGURE 2.3   A woven fabric structure.

2.2 KNITTING

Fabric is formed in knitting by the interlooping of loops of yarns such that one loop is passed through another loop. Due to the loop structure, knitted fabric does not have dimensional stability, but at the same time, it has more stretch and recovery properties. This is most useful in producing inner garments and formfitting garments. The knitted fabrics have contributed a share of about 30% of the total textile fabrics production. Knitted fabrics are bulkier and softer compared to woven fabric, and they are finding application in outer garments in a large way. Even a single yarn is sufficient to form a knitted fabric in knitting. In earlier days, knitting was done using hand pins, then frame knitting was developed. Nowadays, sophisticated knitting machines have been developed with higher production rates for producing a variety of knitted garments to cater to the needs of modern society. Figure 2.4 shows the loops of knitted fabric. Figure 2.5 shows a photograph of a knitted fabric structure.

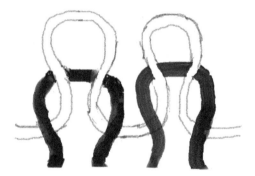

FIGURE 2.4   Loops of knitted fabric.

FIGURE 2.5   Knitted fabric structure.

2.3 NONWOVEN

Nonwoven fabrics are formed directly from the fibres by arranging the fibres in a sheet form and binding the fibres together either mechanically or otherwise. In this process, the necessity of producing the yarn is eliminated. For the production of nonwoven fabrics, synthetic fibres are largely used, and the use of natural fibres are limited. Nonwoven fabrics are not suitable for clothing purposes, but they have a variety of other applications. Technical textiles produced by nonwoven method are finding applications in hygiene products, masks, carpets, floor coverings. automobiles, filters and so on. Figure 2.6 shows the random fibre arrangement in nonwoven fabrics. Figure 2.7 shows a nonwoven fabric.

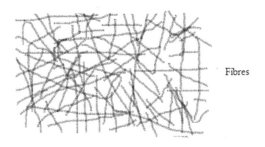

Fibres

FIGURE 2.6   Schematic of the fibre arrangement in nonwoven structure.

FIGURE 2.7   Nonwoven fabric.

Weaving

3 Yarn Preparation for Weaving

The yarn manufactured in the spinning machine contains thick and thin places and objectionable faults. But for satisfactory weaving, the yarn should be free from thick and thin places and objectionable faults. Moreover, the packages from a spinning machine are small in the form of cops which usually weigh about 80g to 100g. To transport the yarn from spinning mills to weaving mills, the smaller packages have to be converted to larger packages. Moreover, the cops have to be emptied for reuse in the spinning section. All these parameters require a process called winding.

3.1 WINDING

Winding is the first process in yarn preparation for weaving. Winding is basically transforming the yarn from small package to another larger package and clearing the yarn from its defects. The warp yarn has to withstand many load cycles during the weaving process, and any defect in the yarn may cause a breakage and causes fabric fault or loom stoppage. Hence, it becomes absolutely necessary to clear the yarn from its defects as much as possible. By the winding process, the yarn faults are removed to the maximum extent.

3.1.1 OBJECTIVES OF WINDING

The main objectives of winding are

1. converting the spinning cops into larger packages weighing about 1kg to 1.5kg.
2. removing the thick and thin places, slubs and sloughs present in the yarn.
3. removing the objectionable faults.
4. giving a wax coating when the yarn is used for knitting.

3.1.2 WINDING PROCESS

The winding process gives an opportunity to clear the yarn from its defects. This is a simple process in which the yarn from the spinning cop is rewound onto a new, larger package. During rewinding, the yarn is cut at the faulty places and rejoined either by knotting or splicing. Figure 3.1 shows the working principle of winding schematically.

DOI: 10.1201/9780367853686-4

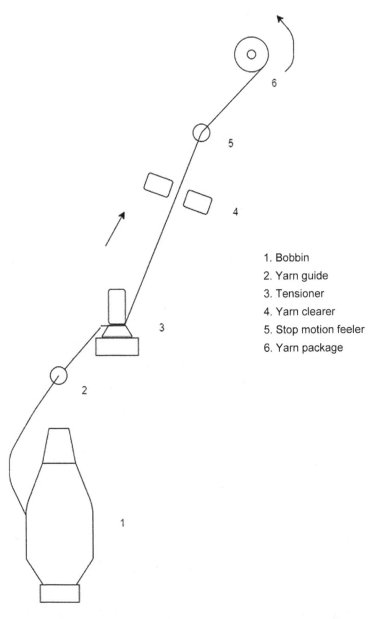

1. Bobbin
2. Yarn guide
3. Tensioner
4. Yarn clearer
5. Stop motion feeler
6. Yarn package

FIGURE 3.1   Schematic diagram of winding.

3.1.3 WINDING MACHINE

The winding machine consists essentially of the following sections:

1. The winding drum
2. Yarn clearers

3. Supply system
4. Stop motion
5. Tensioner

The yarn from the supply package passes through the tension device, yarn clearing unit, stop motion and onto the rotary traverse winding drum. When the winding drum is driven by the motor the take-up package rotates by friction drawing the yarn from the supply package. Whenever a yarn fault, such as thick places, snarls etc., is detected by the clearing unit, the yarn is cut at that place, and the yarn is again rejoined. Thin places also will be removed, because the yarn will break at that place due to tension, and rejoined again. Hence, the majority of the yarn faults will be removed by the winding process.

3.1.3.1 The Winding Drum

The winding drum is a cylindrical grooved metallic or bakelite drum that rotates the take up package by frictional contact. All the high-speed winders adopt the negative friction drive as it is mechanically less complex. The groove in the drum gives traverse movement to the yarn, which lays the yarn throughout the width of the package. This groove arrangement eliminates the necessity for a separate traverse mechanism. The angle of the groove in relation to the drum axis determines the angle of wind on the yarn package. The normal range of angle of wind is from 30° to 55°. As the angle of wind increases the density of the package will degrease. The conicity of the package will vary from 5°57′ to 9°15′ depending on the requirement.

The drums are normally made with 1.0, 1.5, 2.0 and 2.5 scroll. For a 2.5 scroll, the drum makes 2.5 revolutions for one complete traverse, and this corresponds to the minimum angle of wind and maximum density.

3.1.3.2 Yarn Clearers

The process of removing the thin, thick places and objectionable faults is called yarn clearing.

There are two types of yarn clearers:

1. Mechanical yarn clearer
2. Electronic yarn clearer

The mechanical yarn clearer, called slub catchers, consists of two plates separated by a gap as shown in Figure 3.2. This gap can be adjusted according to the requirements by moving the adjustable plate. The yarn passes through the gap between the plates. Whenever a yarn fault having thickness more than the size of the gap enters the gap, it cannot pass through the gap, and hence, the yarn is broken at that place and a knot is put.

The electronic yarn clearer consists of a capacitor having two plates as shown in Figure 3.3. The yarn is passed through the plates. The capacitance of the plates varies whenever the yarn thickness varies. The variation in the capacitance is measured and fed into the controlling unit, and whenever the variation exceeds

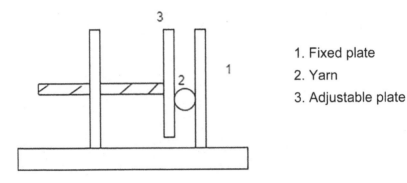

1. Fixed plate
2. Yarn
3. Adjustable plate

FIGURE 3.2   Mechanical yarn clearer.

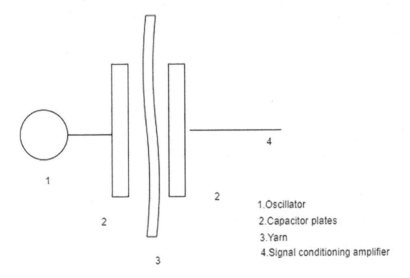

1.Oscillator
2.Capacitor plates
3.Yarn
4.Signal conditioning amplifier

FIGURE 3.3   Electronic yarn clearer.

the predetermined limit, the control unit actuates a cutter, and the cutter cuts the yarn.

The faulty section of the yarn is removed either manually or by air suction; then, the yarn ends are joined by a knot.

3.1.3.3   Supply System

The supply section consists of a bobbin holder and a yarn guide. In automatic winding machines, the bobbin holder will accommodate six or more number of bobbins.

3.1.3.4   Stop Motion

Stop motion is used to stop the winding whenever yarn is broken. This may be mechanical or electronically controlled.

3.1.3.5 Tensioners

A tensioner is provided to give tension to the yarn. There are two types of tensioners: One type is a disc type, and the other is a friction type. In the disc-type tensioner, the yarn passes between two discs. The tension is adjusted by means of adding additional weights on the top disc. In the friction-type tensioner, the tension is adjusted by changing the angle of contact between the yarn and the friction imparting element

3.1.4 YARN FAULTS

Yarn faults are classified into 16 categories according to the length and thickness of the fault. The Uster classmate system is used to classify the yarn faults. The Uster classmate system accounts and classifies the imperfections into 16 groups according to the cross-sectional area and fault length. The four classifications, A, B, C and D, correspond to four fault lengths, namely 0.1cm, 1.0cm, 2.0cm and 4.0cm. A1 to A4 indicates the percentage increase in cross-sectional area from 80% to 400% corresponding to the increase in diameter from 59 % to 153% and similarly for B1 to B4, C1 to C4 and D1 to D4. A1 is the shortest in length and smallest in diameter, and D4 is the longest in length and largest in diameter

3.2 WARPING

Warping is the second process in yarn preparation for weaving. Warping is the process of transferring many yarns from a creel of single end packages into a parallel sheet of yarn wound on a beam called a warper's beam.

3.2.1 NECESSITY FOR WARPING

To weave a cloth of say 140cm width with 30 ends per cm, a total of 4200 warp yarn is required in the beam during weaving. It is not possible to arrange 4200 supply packages or cones in the creel in order to wound the yarn on the weaver's beam. Hence, the total number of yarns are divided into an equal number of smaller sizes. In this case, seven beams are required, each having 600 yarns. Then yarns in the seven beams are put together to get the required 4200 yarn to weave the fabric. This process is known as warping.

3.2.2 WARPING MACHINE

A warping machine consists of a creel system to keep the supply packages, tensioners to impart the necessary tension to the yarn, stop motion to stop the machine whenever yarn is broken and an expanding zigzag comb that can be used to adjust the width of the yarn sheet to the required level. Figure 3.4 shows the schematic diagram of a warping machine. Figure 3.4a shows warping machine creel. Figure 3.4b shows yarn passage in warping machine creel. Figure 3.4c shows the warping machine.

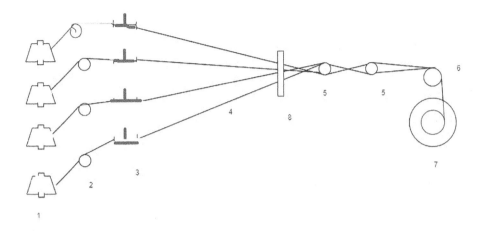

1. Supply package
2. Yarn guide
3. Tensioners
4. Yarn
5. Lease rods
6. Guide roller
7. Warper's beam
8. Expanding zigzag comb

FIGURE 3.4   Working principle of warping.

FIGURE 3.4(A)   Warping machine creel.

FIGURE 3.4(B) Yarn passage in warping machine creel.

FIGURE 3.4(C) Warping machine.

3.3   SIZING

During weaving, due to shed formation, the warp yarn is subjected to stresses and abrasion against adjacent yarn and metal surfaces. Applying a coating of a polymeric film–forming agent on the warp yarn enables the yarn to withstand the stresses to which it is subjected during weaving. This is called slashing or sizing. The following benefits are obtained due to sizing.

3.3.1   BENEFITS OF SIZING

1. The strength of the yarn is increased.
2. Yarn hairiness that would create a problem during weaving is reduced.
3. Abrasion resistance of the yarn is increased.
4. The liberation of fluff and fly during weaving is reduced.
5. Sizing keeps the slack and broken filaments together in low twist yarns.

3.3.2   CHARACTERISTICS OF SIZE FILM

1. The size film should be flexible so that it will not affect the flexibility of the warp yarn.
2. It should not be brittle.
3. The size coating should not be a permanent one. Its purpose will be over once weaving is completed. Hence, size materials should be able to be removed during the desizing process.
4. The size film must coat the yarn surface without excessive penetration into the yarn, because complete desizing will not be possible if the size material is penetrated deeply into the yarn.

3.3.3   SIZING TERMS

The following terms are used in sizing:

Size concentration:
>   The mass of oven dry solid size material in size paste expressed in percentage is called a size concentration.

Size take-up or add-on:
>   The mass of size paste taken up per unit weight of oven-dry unsized yarn expressed in a percentage is called size take-up.

Size percentage:
>   The mass of oven-dry size per unit weight of oven-dry unsized yarn.

3.3.4   OPTIMUM LEVEL OF SIZE ADD-ON

The size add-on percentage should be optimal to give minimum warp breakage during weaving. Excessive size makes the yarn stiff and less extensible. An inadequate size will result in insufficient strength and smoothness.

3.3.5   Sizing Machine

The major parts of sizing machine are the creel, size box, drying unit and beaming. The schematic diagram of a sizing machine is shown in Figure 3.5

3.3.5.1   Creel

Creel is designed in such a way that the necessary number of warp beams are arranged in two rows, one over the other, required to make up the total number of warp yarns in sized beam. Figure 3.5a shows sizing machine creel to accommodate warp beams.

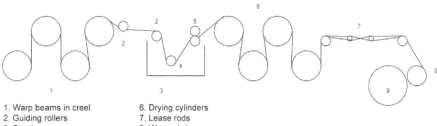

1. Warp beams in creel   6. Drying cylinders
2. Guiding rollers   7. Lease rods
3. Size box   8. Weaver's beam
4. Impregnating roller   9. Winding drum
5. Squeezing rollers

FIGURE 3.5   Working principle of a sizing machine.

FIGURE 3.5(A)   Sizing machine creel.

3.3.5.2   Size Box

Size box is the most important section of sizing machine. During sizing the sheet of yarn is passed under impregnating rolls in the size box containing size mix in hot condition. The amount of size pick-up depends on the sizing condition like size-mix concentration, speed, type of yarn and so on. After impregnating the yarn is passed through squeezing rollers that will squeeze the excess size in the yarn.

3.3.5.3   Drying Unit

The drying unit consists of number of cylinders arranged in two rows one over the other and heated with steam. The yarn impregnated with size is passed over the heated surfaces of the cylinders, which results in drying up of the yarn. Figure 3.5(b) shows drying cylinder of sizing machine.

3.3.5.4   Beaming

The dried sheet of yarn is passed through a pair of lease rods, which will separate the sticky yarns into individual yarn. Then the yarn is wound on the weaver's beam. This is called as beaming. Figure 3.5c shows the beaming unit of sizing machine.

3.3.6   PREPARATION OF SIZE MIX OR PASTE

Aqueous sizes are solutions of one or more adhesives, a lubricant and one or more additives. Generally, aqueous sizes are prepared by cooking size ingredients in size cookers with necessary water and temperature.

FIGURE 3.5(B)   Drying cylinder of sizing machine.

FIGURE 3.5(C) Beaming units of sizing machine.

The ingredients of size mix include the following:

1. Adhesives: Adhesives are usually natural starches such as sago, corn, potato, wheat, modified starches and synthetic sizes, such as polyvinyl alcohol (PVA) and acrylic copolymers.
2. Lubricants The most common lubricants are tallow, spermaceti or paraffin wax Lubricants will be added normally 5% to 10% weight of adhesives. The purpose of a lubricant is to reduce fibre-to-yarn friction during weaving. The reduction of yarn-to-yarn friction facilitates the beat-up of the weft.
3. Additives: Additives are defoamers, emulsifiers, deliquescent or antiseptics.

4 Weaving—Basic Mechanisms

Weaving is a process in which the lengthwise yarns, called warp, and width-wise yarns, called weft, are interlaced. The interlacement of warp and wept yarns results in a material called fabric having tensile strength, flexibility and other essential properties. Each thread or yarn in a warp is called an end, and each thread in the weft is called a pick. The American term for weft is *filling.*

The following six basic mechanisms are essential for continuous weaving:

1. Shedding
2. Weft insertion
3. Beat-up
4. Warp let-off
5. Fabric take-up
6. Stop motions

Shedding, picking/weft insertion and beat-up are called main or primary motions, and warp let-off and fabric take-up are called secondary motions. There are three other motions, namely warp protector, warp stop and weft stop motions, necessary for good weaving. These motions are called auxiliary motions. Figure 4.1 shows the schematics of weaving.

4.1 SHEDDING

Shedding is the process of separating the sheet of warp into two layers one at the top and another at bottom making a V-shape opening as shown in Figure 4.2. This opening is called a shed. Figure 4.2 shows the geometry of warp shed.

The shed height H is given by

$$H = B \tan A,$$

where B is the distance between the heald eye and the fell of the cloth and A is the angle between warp shed and fabric plane. It is advantageous to have a small H in order to reduce the stress on the warp. The magnitude of H is determined by the weft insertion device. Each warp yarn drawn through the eye of a thin metallic bar called a heald is guided by that heald. Healds that guide the warp yarns with the same pattern are attached to the same heald frame. When a heald frame moves up or down, all the yarns attached to that heald frame moves up or down together along with that heald frame. Changing the order of lifting the heald frame will result in a different weaving pattern.

DOI: 10.1201/9780367853686-5

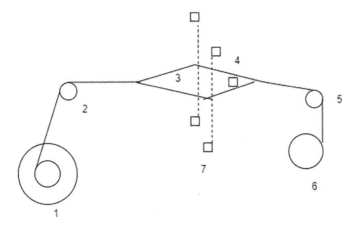

1. Weaver's beam   4. Filling yarn insertion
2. Back rest   5. Fabric guide roller
3. Warp shed   6. Fabric take-up roller
                          7. Heald frame

FIGURE 4.1   Schematics of weaving.

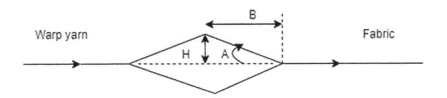

H Shed height
B Distance between heald eye and cloth fell
A Angle between warp yarn and fabric plane

FIGURE 4.2   Warp shed geometry.

4.2   WEFT INSERTION

Weft insertion is the second important motion after shedding in the weaving process. Weft insertion is the process of inserting the weft yarn in between the top and bottom layers of warp shed. This can be achieved by means of a fly shuttle in shuttle weaving and by means of a projectile, rapier and a jet of air or water in shuttleless weaving. Figure 4.3 shows the different weft insertion methods in weaving machines.

4.3   BEAT-UP

Beat-up is the process of pushing the newly inserted weft yarn into the fell of the cloth by means of reed. Reed is a rectangular closed comb of flat metal strips placed

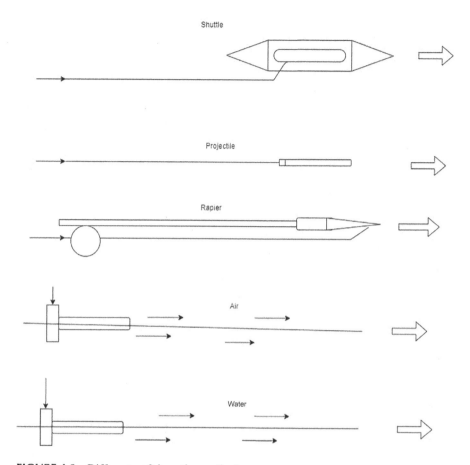

FIGURE 4.3   Different weft insertion methods.

closely with small spacing. This. spacing corresponds to the spacing between the warp yarns. The spaces between two metal strips are called dents. The fell of the cloth is the meeting point of the warp yarn and fabric. Figure 4.4 shows the beat-up process in a loom

Beating up of weft yarn requires a considerable amount of energy. At the final stage of beat-up, the bending of warp and weft yarn takes place due to crimp interchange. To overcome the frictional reactions, the pushing of the weft yarn to the fell of the cloth is done in a harsh manner that gives the name beat-up. During beat-up as the weft yarn is being pushed to the fell of the cloth warp tension increases and the fabric tension decreases.

4.4   WARP LET-OFF

A warp let-off mechanism releases the warp yarn from the sized warp beam at uniform tension as the weaving proceeds. The let-off mechanism controls tension in the warp yarn by controlling the rate of flow of warp yarn. The warp beam diminishes

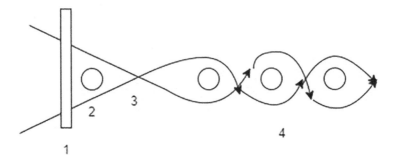

1. Reed

2. Newly inserted weft

3. Fell of the cloth

4. Fabric

FIGURE 4.4   Beat-up.

in diameter as the weaving continues, necessitating a gradual increase in the angular movement of the beam in order to maintain a constant flow of warp.

An increase in warp tension decreases warp crimp and increases weft crimp. Crimp is defined as the ratio between extra length of yarn present to length of fabric where extra length of yarn is the length of yarn present in the sample minus the sample length. The crimp ratio of warp and weft crimp determines the fabric quality. Crimp affects the weight, flexibility and thickness of fabric. Hence, maintaining a uniform warp tension throughout weaving is very important to maintain fabric quality and shrinkage level.

Let-off mechanisms are classified into positive and negative let-off mechanisms. In the negative let-off mechanism, the tension in the warp yarn is used to release the warp yarn from the warp beam. When the tension in the warp yarn exceeds the frictional forces in the let-off mechanism, the beam rotates, and warp yarn is released. Negative let-off mechanism are used in nonautomatic looms. Heavier fabrics with high picks per cm are difficult to weave with a negative let-off mechanism. Figure 4.5 illustrates the principle of a negative let-off mechanism. The frictional force can be increased or decreased by moving the dead-weight away or toward the fulcrum point.

In a positive let-off mechanism, the release of warp yarn is controlled by controlling the rotation of the warp beam positively by a separate mechanism. The mechanism may be either mechanical or electronic. In a positive mechanical let-off motion, the tension variation in the warp is sensed by a lever, and this is fed in to the let-off mechanism through a ratchet wheel, which, in turn, will rotate the beam accordingly. A few examples of positive mechanical let-off motions

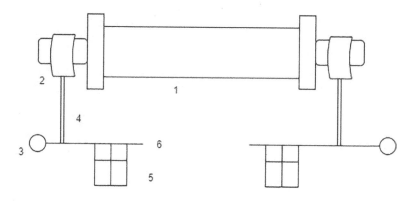

1. Beam
2. Friction element
3. Fulcrum
4. Chain
5. Dead-weight
6. Lever

FIGURE 4.5 Negative let-off mechanism.

are the Bartlett let-off motion, the Ruti let-off motion and the Sulzer Ruti let-off motion. In electronic let-off, a servomotor is used to rotate the beam to the required extent.

4.5 FABRIC TAKE-UP

Fabric take-up is an important function in a loom because it controls the quality of the fabric in terms of picks per unit length. Uniform fabric take-up will result in uniform pick density. There are two types of take-up mechanisms: (1) mechanical and (2) electronic.

4.5.1 MECHANICAL TAKE-UP

The rotation of the fabric roller is determined by the change and standard wheels in the fabric take-up gear. The required amount of pick density is achieved by changing the change and standard wheels in the fabric take-up gear in a conventional shuttle loom. In automatic shuttle looms the standard and change wheels are dispensed with instead levers are used to change pick density. The schematic diagram of a mechanical take-up mechanism used in an automatic shuttle loom is given in Figure 4.6.

A double-through take-up cam 2 is mounted on the picking shaft 1, and as the cam rotates, it oscillates the follower bracket 4. This motion is transmitted to the actuating lever 6 loosely mounted on the ratchet wheel stud. The left end of the connecting rod 5 rests on the slot of the actuating lever. The up and down adjustment of the connecting rod in the slot determines the pick density. An upward adjustment results in more angular movement of the ratchet pawl and decreases the pick density. A downward

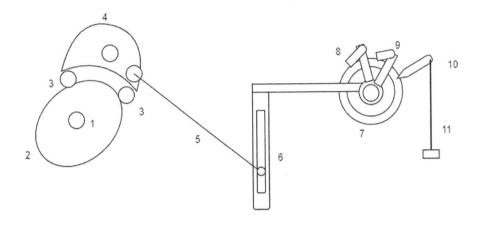

1. Picking shaft	7. Ratchet wheel
2. Take-up cam	8. Driving pawl
3. Followers	9. Stop pawl
4. Follower bracket	10. Catching pawl
5. Connection rod	11. Pawl release bracket
6. Actuating lever	

FIGURE 4.6   Lakshmi Ruti C mechanical take-up motion.

adjustment results in increased pick density. No change wheels are required to alter the pick density.

4.5.2 ELECTRONIC TAKE-UP

The rotation of the fabric roller is done by a servomotor. The weft density can be altered by changing the frequency of the servomotor. The schematic diagram of an electronic take-up mechanism used in an air-jet loom is given in Figure 4.7. The fabric is drawn over a spreading roll then under the fabric roll before it is wound on a cloth roll. A press roller is used to prevent the fabric from slipping back.

4.6 STOP MOTIONS

Warp and weft yarns are liable to break during weaving operation. During weaving, the warp and weft yarn is subjected to tension, and whenever the tension in the yarn exceeds the strength of the yarn, the yarn breaks. Yarn faults, such as weak places, snarls, slubs and so on, present in the yarn also leads to yarn breaks. Hence, it becomes necessary to stop the loom as soon as either the warp or weft yarn breaks. This stopping mechanism may be mechanical or electrical.

The mechanisms used to stop the loom whenever a weft breaks or weft exhausts are called weft stop motions. Similarly, mechanisms used to stop the loom whenever a warp yarn breaks are called warp stop motions.

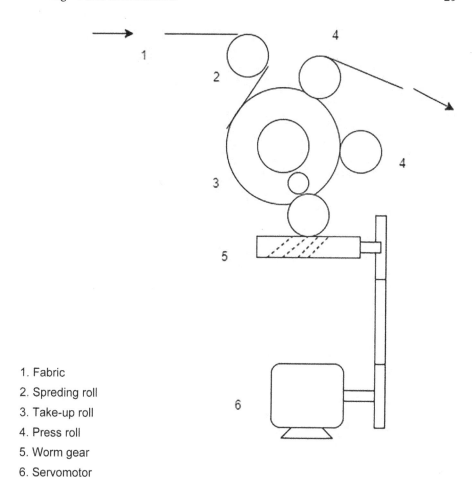

1. Fabric
2. Spreding roll
3. Take-up roll
4. Press roll
5. Worm gear
6. Servomotor

Text

FIGURE 4.7 Electronic fabric take-up.

5 Shedding

Shedding is an important motion in weaving operation. Without shedding weaving cannot be done. All designs are created only by a shedding motion. The properties of the final fabrics, to a larger extent, depend on the shedding. There are three types of shedding systems available for the formation of warp shed in weaving machines:

1. Tappet or cam shedding
2. Dobby shedding
3. Jacquard shedding

The cam and dobby shedding mechanisms control the heald frames and each heald frame controls a group of warp threads. The maximum no of heald frames possible in a cam shedding is eight. Hence, only weave patterns having a repeat of less than eight warp threads are possible to weave in cam shedding. This limitation is due to the design of the cam. Up to 24 heald frames can be accommodated in a dobby shedding. Therefore, weave patterns having a repeat of up to 24 warp threads can be woven in dobby shedding. Warp threads are controlled individually in jacquard shedding. The limitation for the number of warp threads in a repeat in jacquard shedding is the capacity of the jacquard, which is usually 400,800 or 1600hooks.

5.1 TAPPET OR CAM SHEDDING

Cam mechanisms are simple and easy to operate. A majority of weave pattern commonly used can be produced using cam shedding. The disadvantage of cam shedding is that whenever the weave pattern is changed, it becomes necessary to rearrange the cams, which is a time-consuming process.

A cam is a disk that transforms the rotational motion into the reciprocating motion of the follower. The transfer is done by means of the cam's edge, in the ease of a negative cam, and by means of a groove cut in the side surface of the cam, in the case of a positive cam. Figure 5.1 illustrates negative and positive cams.

5.1.1 NEGATIVE CAM SHEDDING

The negative cam mechanism acts only in one direction. It either raises or lowers the heald frames. A reversing mechanism is necessary to return the heald frames. A spring reversing motion is used in most cases. Figure 5.2 illustrates the action of negative cam shedding. As cam 1 rotates, it presses follower 2 downwards. This will make traddle 3 to pull heald frames 5 downwards against the action of spring 6. As the cam rotates further, the edge of the cam moves up paving the way for the spring to lift the heald frames and the traddle.

DOI: 10.1201/9780367853686-6

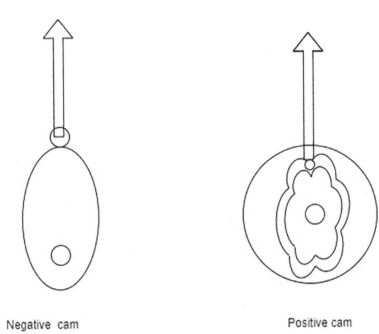

Negative cam   Positive cam

FIGURE 5.1   Negative and positive cams.

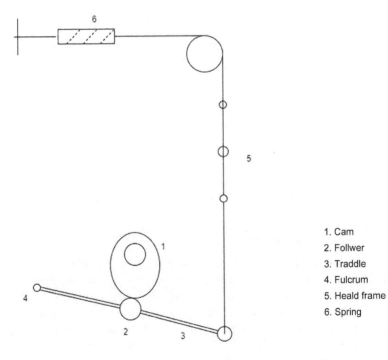

1. Cam
2. Follwer
3. Traddle
4. Fulcrum
5. Heald frame
6. Spring

FIGURE 5.2   Negative cam shedding.

5.1.2 Positive Cam Shedding

The heald frames are raised and lowered by cams in the positive cam shedding. There are two types of positive cams. In the first, a roller follows a groove cut in the face of the cam. An 'L'-shaped lever is attached to the roller, and when the roller moves up and down as the cam rotates the other end of the lever moves back and forth in a horizontal direction. This motion is carried to the heald frames through various levers.

In the second type of positive cam shedding a pair of matched cams are used for each heald frame. The two rollers, which are in contact with cam faces, oscillate the lever to which they are attached. This oscillation is converted into an up-and-down movement of heald frames through levers. Figure 5.3 shows a positive cam system with two negative cams.

5.2 DOBBY SHEDDING

Whenever the number of heald frames exceeds 8 a dobby mechanism is necessary for shedding. A dobby consists of two parts. The first part is a lifting mechanism to lift the heald frames, and the second part is a selection mechanism to select the healds to be lifted. There are two types of dobbies: (1) negative dobbies and (2) positive dobbies.

5.2.1 Negative Dobby

In a negative dobby, the heald frames are lifted by the dobby and lowered by spring motion. Figure 5.4 shows a schematic of a double lift negative dobby. Two knives, 3 and 4, with a pair of hooks, 5 and 6, are used to lift the heald frames 9 through baulk 1 and jack 8 while the heald frames are lowered by springs 10. Whenever peg 14 in the design cylinder 13 comes vertical under selection lever, the peg lifts selection

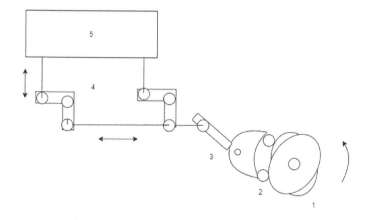

1. Matching cams
2. Follower
3. Oscillating lever
4. Link mechanism
5. Heald frame

FIGURE 5.3 Positive cam system with two negative matching cams.

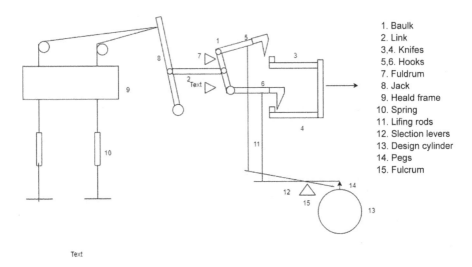

1. Baulk
2. Link
3,4. Knifes
5,6. Hooks
7. Fuldrum
8. Jack
9. Heald frame
10. Spring
11. Lifing rods
12. Slection levers
13. Design cylinder
14. Pegs
15. Fulcrum

FIGURE 5.4   Schematic of a double-lift negative dobby.

lever 12, and the other end of the selection lever is lowered. This makes the lifting rod 11 come down slightly, which makes the lower hook also come down in the way of the lower knife. When the lower knife moves forward, this will pull the lower hook forward, and as a result, the baulk and jack are made to make an angular movement. This angular movement lifts the heald frames through strings. The pegs are pegged in the lag, and all the lags are joined to make a chain. Each lag corresponds to two picks.

In modern machines, as an alternative to lag-and-peg chains, punched paper or plastic pattern cards are used. A punched hole in the paper corresponds to a peg in the lag and this causes the corresponding heald frame to lift.

5.2.2 Positive Dobby

The heald frames are lifted and lowered both by the dobby mechanism. The need for the springs is eliminated. As a result, positive dobbies are capable of running at higher speeds. Rotary dobbies are good examples for positive dobbies. Figure 5.5 shows the working of rotary dobby. As cam 2 rotates along with cam shaft 1, it produces a rocking movement in rocking arm 4. As the rocking arm makes an angular movement, the other end makes a traverse movement. This is used to lift and lower the heald frames through a linking mechanism.

5.3 JACQUARD SHEDDING

The first jacquard machine was invented by Joseph Marie Jacquard (1752–1834). Jacquard shedding offers individual control of warp threads, and as a result, heald shafts are not required. They are simple in principle and construction, with unlimited patterning possibilities. Jacquard machines contain many parts, which are difficult

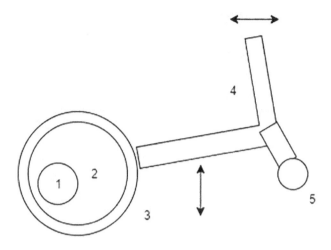

1.Cam shaft

2.Cam

3.Follower

4.Rocking arm

5.Fulcrum

FIGURE 5.5   Rotary dobby.

to maintain. Figure 5.6 shows the working principle of single-lift single-cylinder jacquard machine.

A jacquard machine consists of two parts: a selection mechanism and a lifting mechanism. The selection mechanism consists of needles 5, 6; spring box 4; pattern cards 8; and cylinder 7. The lifting mechanism consists of hooks 2, 3; knife 1; grate 9; and springs 11. The fabric design is punched in pattern cards, and each card represents one pick in the weave. There is one needle and one hook for every warp thread in the repeat. Pattern card 6 is presented to cylinder 7 in such a way that each card fits one side of the pattern cylinder. For every pick, the cylinder rotates and presses against the needles. The coil springs in spring box 4 also press the needles toward the cylinder. Suppose that a hole is there in the punched card against needle 5 and the needle will go inside the hole positioning hook 2 against knife 1. When the knife makes an upward movement, hook 1 will be lifted, making the warp thread attached to the hook to form the upper shed. If there is no hole in the card, as for the case of needle 6, the needle will be moved horizontally, making hook 3 deflect away from the knife. The knife will not lift hook 3, making the warp thread attached with hook 3 form the bottom shed. After each pick, the cylinder is moved away from the needles, rotated and presented against the needles for the next pick.

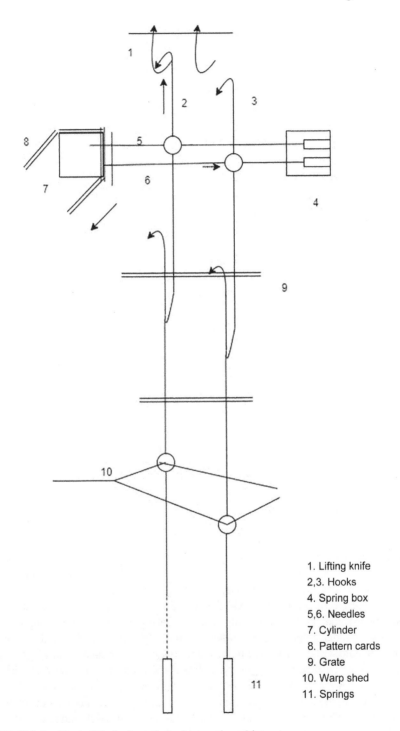

FIGURE 5.6 Single-lift single-cylinder jacquard machine.

1. Lifting knife
2,3. Hooks
4. Spring box
5,6. Needles
7. Cylinder
8. Pattern cards
9. Grate
10. Warp shed
11. Springs

The capacity of jacquard machine is indicated by the number of hooks, and each hook will control one warp thread. For example a 600-hook jacquard will control 600 warp threads. Single-lift jacquard machines are slow in operation. Therefore, modern weaving machines employ double-lift jacquard machines, which can run at higher speeds. A double-lift jacquard has two sets of knives that move up and down opposite to each other in a two-pick cycle. Electronically controlled double-lift jacquards are used in shuttleless weaving machines.

6 Shuttle Weaving

A shuttle loom uses a shuttle to carry the weft yarn. The shuttle inserts the weft yarn in the shed during its traverse back and forth across the loom width. On both sides of the loom, picking mechanisms are fitted, which will propel the shuttle to the other end. The shuttle can be made out of wood or plastic. Figure 6.1 shows a shuttle used in conventional looms. A pirn is a small piece of tube made out of plastic or wood on which the weft yarn is wound. The size of the shuttle will vary according to the loom and normally it will be able to hold various pirns of different sizes. The size of a pirn will vary from 15cm to 25cm according to the requirement.

6.1 CONVENTIONAL LOOMS

In conventional looms as the weaving proceeds, the weft yarn is unwound from the pirn and inserted in the warp shed. Since the weft yarn package has a limited quantity of yarn, it will run out after a limited time. Usually after every 5 to 6 minutes, the loom has to be stopped, and the pirn has to be changed manually. Figure 6.2 shows the schematic of a shuttle loom. Splitting the warp yarn into the top and bottom layers is called shedding. Propelling the shuttle into the warp shed with a speed is called picking. On each side of the loom, there is a picking stick which will hit the shuttle to fly to the other side of the loom. The shuttle will fly at a speed of approximately 15m/sec. On reaching the other side the shuttle will be stopped by shuttle checking system.

There are different mechanisms for shuttle picking. Figure 6.3 shows a cone-under-pick mechanism used in conventional looms. A cam is fixed in the cam shaft, which is continuously rotating. The cam hits the cone of the picking shaft, which turns and pulls the lug strap. The picking stick attached with the lug strap is also pulled, and it hits the shuttle through the picker, making the shuttle to fly to the other side of the loom. If it is assumed that the speed of the shuttle is 15m/sec and for a loom having a width of 150cm, the shuttle will take 0.1 sec to reach the other side.

The shuttle travels on the race board while the bottom layer of the warp yarn is between the shuttle and the race board. The reed guides the shuttle, and the race board supports the shuttle.

The reed and the race board are assembled on the two oscillating arms, which is called a sley. The sley oscillates back and forth by means of a crank mechanism attached with the main shaft of the loom as shown in Figure 6.4. During its forward movement, the sley pushes the inserted weft yarn to the fell of the cloth. This is called beat-up. The meeting point of the warp yarn and the newly formed cloth is called the fell of the cloth.

DOI: 10.1201/9780367853686-7

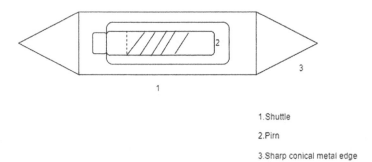

1.Shuttle

2.Pirn

3.Sharp conical metal edge

FIGURE 6.1   Shuttle.

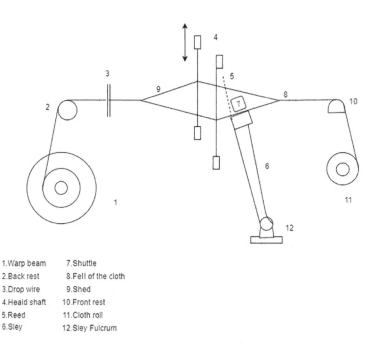

1.Warp beam	7.Shuttle
2.Back rest	8.Fell of the cloth
3.Drop wire	9.Shed
4.Heald shaft	10.Front rest
5.Reed	11.Cloth roll
6.Sley	12.Sley Fulcrum

FIGURE 6.2   Schematic of a shuttle loom.

6.2   TIMING DIAGRAM OF SHUTTLE LOOM

All the basic motions namely shedding, picking and beat-up are synchronized in a loom for its proper working. Taking the revolution of the crank-shaft as 360° and the beat-up position as 0° the start and end of all the motions are indicated in a circle. This is called a timing diagram of the loom. The timing diagram of a shuttle loom is demonstrated in the Figure 6.5. The timing diagram may be different for different looms. The beat-up position that is the furthermost position of the sley represents 0°. Picking starts 80° from the beat—up and lasts for 30°. The shed is fully open from 30° to 150°. Shuttle checking takes place from 260° to 290°.

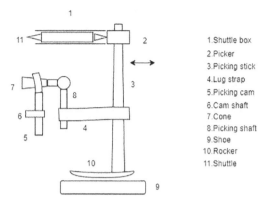

FIGURE 6.3   Schematic of a cone-under-picking mechanism.

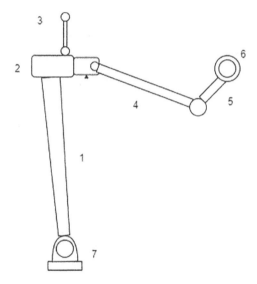

FIGURE 6.4   Schematic of sley and beat-up mechanism.

6.3   AUTOMATIC LOOMS

In a conventional loom, as soon as the weft yarn in the pirn exhausts, the loom has to be stopped, the exhausted pirn should be replaced with a full pirn manually and the loom needs to be started again. This will reduce the efficiency of the loom and involves manpower. A starting mark of one or two picks will be left in the fabric. To make the loom continuously run without stoppage for weft replacement, pirn-changing mechanisms have been developed. Pirn-change mechanisms replace the exhausted pirn with a full pirn while the loom is running. The looms fitted with pirn-change mechanisms are called auto looms. Figure 6.6 shows the working of a pirn-change mechanism.

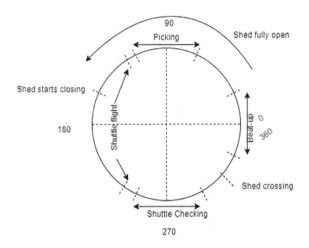

FIGURE 6.5 Timing diagram of shuttle looms.

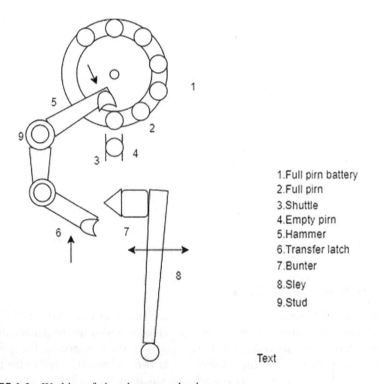

1.Full pirn battery
2.Full pirn
3.Shuttle
4.Empty pirn
5.Hammer
6.Transfer latch
7.Bunter
8.Sley
9.Stud

Text

FIGURE 6.6 Working of pirn-change mechanism.

Following the indication from the weft feeler for the exhaustion of weft, transfer latch 6 is lifted so that it positions itself against approaching bunter 7. As sley 8 moves forward, the bunter hits the transfer latch, which results in hammer 5 hitting the full pirn kept in battery 1 downwards, ejecting the empty pirn from the shuttle. The full pirn will position itself in the shuttle, and weaving continues without stoppage. All these things will happen in fraction of a second during the furthermost position of the sley.

7 Shuttleless Weaving

7.1 LIMITATIONS OF SHUTTLE WEAVING

Even though shuttle weaving is the most common and simple method of production of fabrics, for many centuries, it has its own limitations. The limitations of the shuttle weaving are:

1. Shuttle weaving requires frequent replenishment of weft packages due to its smaller size, which increases the workload of the weaver.
2. Because the shuttle has to be propelled back and forth, it requires a heavy mechanism, resulting in the consumption of spare parts.
3. A separate pirn winding process is necessary to prepare the weft packages.
4. A few layers of weft yarn are always left in the pirn after every full pirn used in the weaving. This needs to be cleaned before the pirn is used again for winding, resulting in wastage of weft yarn to a certain extent.
5. The speed of the loom cannot be increased beyond a certain limit say 250 rpm due to the heavy moving parts of the loom and mass of the shuttle.

All the limitations listed above are mainly due to the method of weft insertion by the shuttle system.

7.2 PRINCIPLES OF SHUTTLELESS WEFT INSERTION

To overcome the limitations of shuttle weaving, new weft insertion systems have been developed. These are called shuttleless weaving. Mainly four shuttleless weft insertion principles are commonly used:

1. Projectile weft insertion
2. Rapier weft insertion
3. Air-jet weft insertion
4. Water-jet weft insertion

7.3 WEFT ACCUMULATORS

Even though the weft packages are dispensed with in a shuttleless system, it becomes necessary to measure the weft yarn required for each pick and kept ready for insertion. For this purpose, weft accumulators or feeders are used in shuttleless weaving. The main function of the weft accumulators is to unwind the yarn from the larger yarn package and supply to the weaving machine smoothly at constant and proper tension. The weft insertion velocities for various insertion systems are given in Figure 7.1

DOI: 10.1201/9780367853686-8

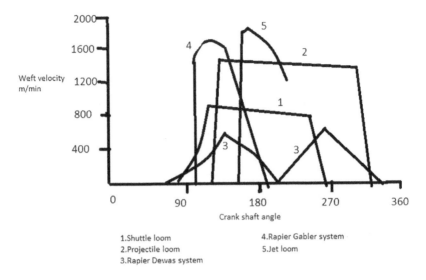

FIGURE 7.1   Weft insertion velocity for various insertion systems.

There are two types of weft accumulators:

1. Drum storage accumulator
2. Loop storage accumulator

7.3.1 DRUM STORAGE ACCUMULATOR

The weft yarn from the supply package is drawn at a constant rate and wound on a highly polished metal cylindrical body. During weft insertion, the coils are free to be pulled off. A tensioner is provided between the drum and the supply package. A stopper is provided between the weft insertion element and the drum to prevent extra supply of weft yarn over than necessary for one pick. Figure 7.2 shows the schematic of drum storage–type weft accumulator. Figure 7.2a shows a drum storage accumulator.

When yarn guide 7 rotates around storage drum 4, yarn is taken from supply package 1 through tensioner 3 and wound on the storage drum. As the yarn is withdrawn during picking, the tension in the yarn is controlled by the braking device. Different types of braking devices, such as bristle, metal lamella, flex brake and endless beryllium copper tensioning strips, are used.

Advantages of weft accumulator are

1. a decrease in the average tension during weft insertion.
2. fewer weft breakages.
3. equalization of unwinding yarn tension due to the diminishing diameter of weft supply package.

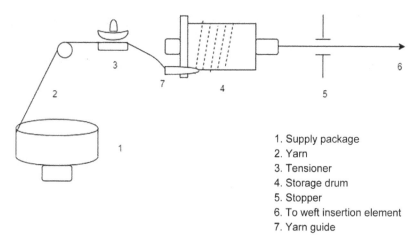

1. Supply package
2. Yarn
3. Tensioner
4. Storage drum
5. Stopper
6. To weft insertion element
7. Yarn guide

FIGURE 7.2 Schematic of weft accumulator (drum storage type).

FIGURE 7.2(A) Drum storage accumulator.

4. the equalization of uneven unwinding characteristic of weft package of different types.

7.3.2 LOOP STORAGE ACCUMULATOR

In a loop storage accumulator, a measuring roller draws the weft from the supply package at a constant rate, and this yarn is stored in the form of a loop in a tube with the help of suction. Figure 7.3 shows a loop storage type accumulator. This type of accumulator is not used anymore.

7.4   SELVEDGES

The warp-way strips that form the edges of the fabric are called selvedges. Selvedges provide strength to fabric for the subsequent handling. It binds the extreme outer ends of warp yarn with the weft to prevent the cloth fraying. A firm, smooth and perfect selvedge is necessary for further processing of fabric and for future end use in some cases. The thickness of the selvedge should be the same thickness as the fabric as for as possible. If the selvedges are too thick, they lead to damage during calendaring. There are four main types of selvedges:

1. Conventional or shuttle selvedge
2. Tucked-in selvedge
3. Leno selvedge
4. Fused selvedge

7.4.1   CONVENTIONAL OR SHUTTLE SELVEDGE

In shuttle loom, the shuttle carries the same weft yarn during its return movement without cutting the weft at the edges. Therefore the weft binds the warp yarn at the edges with the body of the fabric. The shuttle selvedges are strong and smooth and look clean and uniform. It is the best selvedge of all other selvedges. Figure 7.4 shows a shuttle selvedge. A few extra warp threads are necessary to produce strong selvedges. The width of the selvedge may vary from 10mm to 15mm.

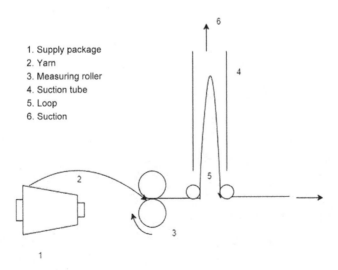

1. Supply package
2. Yarn
3. Measuring roller
4. Suction tube
5. Loop
6. Suction

FIGURE 7.3   Weft accumulator (loop storage type).

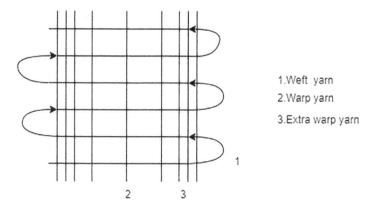

1.Weft yarn

2.Warp yarn

3.Extra warp yarn

FIGURE 7.4   Conventional selvedge.

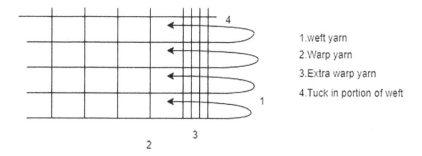

1.weft yarn

2.Warp yarn

3.Extra warp yarn

4.Tuck in portion of weft

FIGURE 7.5   Tucked-in selvedge.

7.4.2   TUCKED-IN SELVEDGE

In the shuttleless looms, both ends of the weft yarn are cut, and therefore, they do not bind the warp threads at the. edges along with the body of the fabric. While cutting the weft at both ends, a length of about 10mm to 15mm is left as fringe. This projecting weft tail is turned and woven back into the body of the fabric by a special mechanism in the form of a hairpin during the next picking. This tucking in of the weft tail at both edges of the fabric leads to the formation of selvedges. This selvedge is called a tucked-in selvedge. In tucked-in selvedge, the density of the weft yarn at the selvedge area is double. Next to conventional selvedge, tucked-in selvedges are good performers. The disadvantage of tucked-in selvedge is its initial cost for the tuck-in mechanism. Almost all types of shuttleless weaving machines use tucked-in mechanisms. Figure 7.5 shows a tucked in selvedge.

7.4.3   LENO SELVEDGE

Leno selvedges are formed by leno design at the edges of the fabric. Fabrics with leno selvedge has fringe edges which are less attractive. However, leno selvedges are most suitable for all shuttleless weaving machines. In half-cross leno selvedge, two

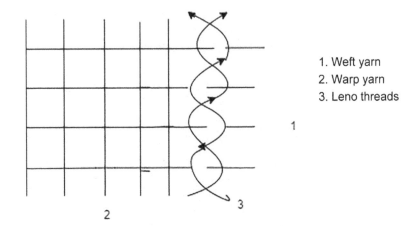

1. Weft yarn
2. Warp yarn
3. Leno threads

FIGURE 7.6   Leno selvedge.

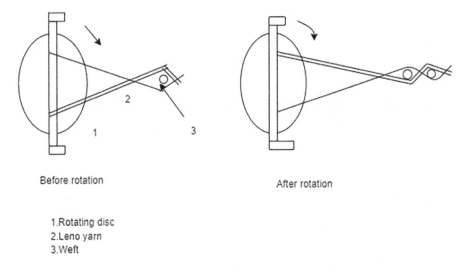

Before rotation

After rotation

1.Rotating disc
2.Leno yarn
3.Weft

FIGURE 7.7   Formation of full-cross leno selvedge.

leno threads run in opposite directions around the standard end, thus binding each pick on both sides of the standard end as shown in Figure 7.6.

A full-cross leno selvedge is made by twisting two leno threads continuously together and inserting the weft between them. A full leno selvedge is formed by a rotating disc, which has two flanged bobbins at the rear of the healds. The sequence of selvedge formations is shown in Figure 7.7

7.4.4   FUSED SELVEDGE

The outer ends of the warp are fused with the weft at the selvedges by means of heaters. This is called a fused selvedge. Fused selvedges can be formed only when weaving thermoplastic filaments yarn such as polyester, polyamide and polyolefin.

8 Projectile Weaving

Projectile weaving machines uses a projectile equipped with a gripper to insert the wept in the warp shed instead of a regular shuttle. The size of the projectile is much smaller, 90mm long, 14mm wide and 6mm thickness. The weight of the projectile is only 40g while the weight of the shuttle is 400g. A newly developed torsion rod picking mechanism is used for picking. Picking always takes place from one side of the machine only. Several projectiles are used, and all of them are returned to the picking side by a conveyor. The projectile is guided in the warp shed by rake-shaped guides. The projectiles are made out of steel or composite. Figure 8.1 illustrates a schematic of a projectile and a gripper, and Figure 8.1a shows a projectile and a gripper.

1.Steel projectile

2.Yarn gripper

FIGURE 8.1   Schematic of a projectile and a gripper.

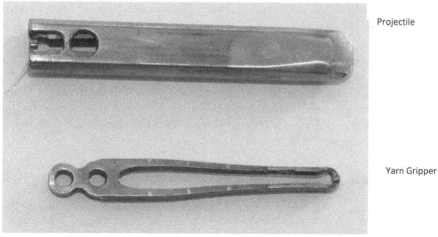

Projectile

Yarn Gripper

FIGURE 8.1(A)   Projectile and gripper.

DOI: 10.1201/9780367853686-9

8.1   PRINCIPLE OF PROJECTILE WEFT INSERTION

There are seven separate operations in the sequence of the projectile weft insertion cycle. Figure 8.2 illustrates all the operations of the projectile weft insertion cycle.

Position A: Projectile 7 moves into the picking position.

Position B: Projectile feeder 5 opens after the weft yarn is gripped by the gripper.

Position C: Picking arm 6 hits the projectile, and the projectile has reached the other side of the machine after inserting the weft in the warp shed.

Position D: The projectile was stopped by projectile brake 9, while weft tensioner 4 stretches the weft, and at the same time, projectile feeder 5 moves close to the cutter.

Position E: Projectile feeder 5 grips the weft yarn.

Position F: The projectile releases the weft yarn; cutters 8 at both sides move forward and cut the weft near the selvedges. The projectile is released and expelled by the projectile brakes. The expelled projectile is carried by the conveyor to the picking side.

Position G: The reed has beaten up the weft yarn. The projectile feeder moves backwards to the feeding position while the weft tensioner takes up the slacken weft caused by the backwards movement of the feeder. Now the system is ready for the next picking cycle.

8.2   WARP LET-OFF MECHANISM

In older versions of projectile weaving machines, warp let-off is controlled mechanically. Figure 8.3a shows a mechanical warp let-off motion. A worm and worm wheel are used to drive the beam. The worm is getting motion from the main shaft through a ratchet arrangement. The ratchet is connected to the whip roll through links. Tension in the warp is monitored and transmitted to the ratchet. Ratchet will rotate more when tension is more and rotate less when tension in warp is less. But in later versions warp let-off is controlled electronically. A back-rest roller serves as an automatic control device. The position of the back-rest roller is sensed by a sensor and transmitted to the control unit. Whenever the back-rest roller moves away from its reference position due to tension variation in the warp, the control unit will speed up or slow down the warp let-off motor. From the motor the motion is transmitted to the warp beam via a worm and gear wheels. Figure 8.3 illustrates an electronically controlled warp let-off motion. Figure 8.3b shows an electronic warp let-off motion.

8.3   SHEDDING MECHANISMS

Projectile weaving machines are equipped with either positive cam or rotary dobby systems.

The rotary dobby can be controlled either mechanically or electronically. A shed-levelling system is used to level the heald shafts while the machine is stopped for

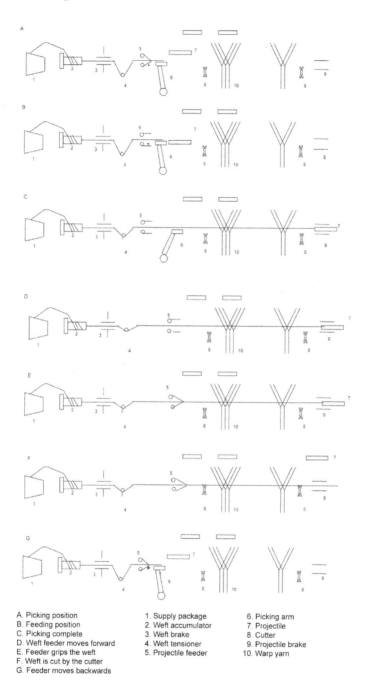

A. Picking position
B. Feeding position
C. Picking complete
D. Weft feeder moves forward
E. Feeder grips the weft
F. Weft is cut by the cutter
G. Feeder moves backwards

1. Supply package
2. Weft accumulator
3. Weft brake
4. Weft tensioner
5. Projectile feeder

6. Picking arm
7. Projectile
8. Cutter
9. Projectile brake
10. Warp yarn

FIGURE 8.2   Projectile weft insertion cycle.

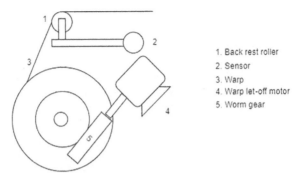

1. Back rest roller
2. Sensor
3. Warp
4. Warp let-off motor
5. Worm gear

FIGURE 8.3   Schematic of electronic warp let-off motion.

FIGURE 8.3(A)   Mechanical warp let-off motion.

FIGURE 8.3(B)   Electronic warp let-off motion.

mending broken warp threads. Figure 8.4 shows a positive cam shedding mechanism. Figure 8.4a shows a positive cam.

When positive cam 1 rotates, it oscillates roller lever 6 horizontally. This horizontal motion is transmitted to the heald shafts through the links as a vertical up and down movement. By adjusting the height of the connecting arm with the roller lever, the shed opening can be adjusted.

8.4 TORSION-ROD PICKING MECHANISM

The unique feature of a projectile weaving machine is its torsion-rod picking mechanism. The energy required for picking is built up and stored in a specially made torsion rod and suddenly the torsion rod is released during picking, making the picking arm to hit the projectile. The projectile is accelerated, and it travels to the other end of the machine through the rake-shaped projectile guides. Brakes in the receiving unit stop the projectile, and it is ejected to the conveyor belt, which carries the projectile to the picking side. Figure 8.5 illustrates the torsion-rod picking mechanism. When cam 7 rotates, it brings roller lever 5 and knee joint 4 into line, thereby twisting torsion rod 8 as shown in Figure 8.5a before picking. Further movement of the cam makes roller lever 5 and knee joint 4 move beyond the straight-line dead point, causing the knee joint to yield as shown in Figure 8.5b after picking. This enables the twisted torsion rod to release. Picking lever 9 makes a sudden angular movement hitting projectile 2 via picking shoe 1. The projectile gets accelerated and travels to the other side at a high velocity. The movement of the torsion rod is cushioned by oil brake 6. Figure 8.5c shows the picture of cam, roller lever and oil brake.

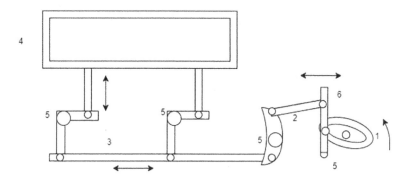

1. Positive cam
2. Connecting arm
3. Link mechanism
4. Heald shafts
5. Fulcrum
6. Roller levers

FIGURE 8.4 Positive cam shedding mechanism.

FIGURE 8.4(A)   Positive cam.

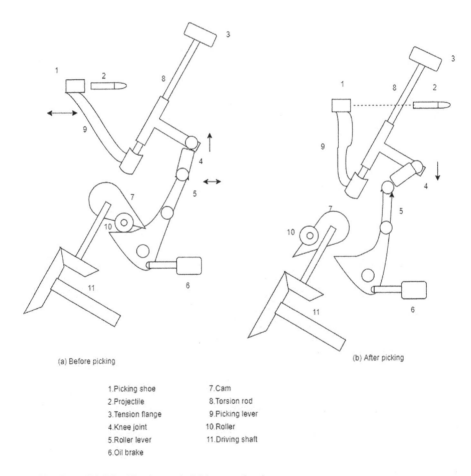

1.Picking shoe	7.Cam
2.Projectile	8.Torsion rod
3.Tension flange	9.Picking lever
4.Knee joint	10.Roller
5.Roller lever	11.Driving shaft
6.Oil brake	

FIGURE 8.5(A),(B)   Torsion-rod picking mechanism.

FIGURE 8.5(C) Picture of cam, roller lever and oil brake.

8.5 BEAT-UP

The beat-up mechanism in a projectile weaving machine is different from the conventional crank and connecting rod system. The sley is positively rocked about its centre by means of matched cams. The sley arm is very short compared to the sley arm of the shuttle looms, and its mass also is much less, about 16kg for a 220cm loom. The sley carries staggered projectile guides throughout the loom width to guide the projectile. As a result, the projectile travels without toughing either the warp thread or the race board during its movement. During beat-up, the projectile guides come below the warp threads. Figure 8.6 shows the matched-cam beat-up mechanism of projectile weaving machine and Figure 8.7 shows projectile loom sley.

Saddle 1, carrying two antifriction bowls, is attached to sley 3. When matched cam 4 rotates, it rocks the saddle positively, which, in turn, makes the sley to oscillate. Reed 5, attached with the sley, moves forwards and backwards, thereby beating the newly inserted weft into the fell of the cloth. During the forward movement of the sley, projectile guide 6 comes under the warp yarn. Because of the matched cam system, the sley dwell at the back centre can be adjusted so that during the

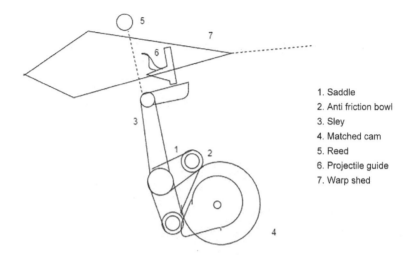

1. Saddle
2. Anti friction bowl
3. Sley
4. Matched cam
5. Reed
6. Projectile guide
7. Warp shed

FIGURE 8.6   Matched-cam beat-up mechanism of projectile weaving machine.

FIGURE 8.7   Projectile loom sley.

entire travelling period of the projectile the sley is in dwell position. The dwell periods vary according to the width of the loom. The sley dwell for 360cm and above machines is from 110° to 340°. For 190cm to 220cm machines, it is from 150° to 340°.

8.6   SALIENT FEATURES OF PROJECTILE WEAVING MACHINES

The projectile weaving machines are very popular and convenient to use. The salient features of projectile weaving machines follow:

1. The picking and projectile receiving units are separated from the moving sley, making the picking mechanism less weight and less complicated.

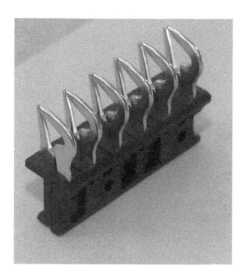

FIGURE 8.8 Projectile guide.

2. Since for movement of the projectile separate guides are provided, warp damage is avoided due to the frictional force between projectile and warp yarn. Figure 8.8 shows a projectile guide.
3. Picking always takes place on one side only.
4. Every weft is cut off on both sides.
5. Tucked-in selvedge mechanisms are provided at both ends. Therefore, selvedges look like a conventional selvedge.
6. More than one fabric can be woven at the same time.
7. The clamping force of the gripper ranges from 600g to 2500g according to its type.
8. The weft insertion rate of projectile weaving machines varies from 500mpm (metres per minute) to 1500mpm according to its width and type.
9. A small warp shed leads to less warp breakage.

9 Rapier Weaving

Rapier weaving machines use a rigid or flexible element called rapier to insert the weft yarn in the warp shed. The rapier head picks up the weft yarn from one side and carries it to the other side of the weaving machine. The rapier returns empty to pick the weft yarn for the next pick, which completes one cycle. The rapiers perform reciprocating motions. Rapier weaving machines are reliable and flexible. A wide range of fabrics can be woven, from lightweight to heavyweight. Rapier weaving machines can weave cotton, wool, silk and synthetic fibres, such as polyester, amongst others.

9.1 TYPES OF RAPIER WEAVING MACHINES

Rapier weaving machines are of two types. Rapier weaving machines that use only one rapier head are called single rapier machines. Rapier weaving machines that use two rapier heads are called double rapier machines. Double rapier machines are further classified as double rigid rapiers and double flexible rapiers.

9.1.1 SINGLE RAPIER MACHINES

A single rigid rapier is used in these machines for weft insertion. The rigid rapier is made of metal or composite with a round cross section. The rapier enters the warp shed from one side of the machine and, on reaching the other side, picks up the tip of the weft yarn and inserts the weft yarn in the warp shed while retracting. Figure 9.1 illustrates the weft insertion cycle of single rapier machines.

In Figure 9.1 at position A, rapier 1 is ready to enter the warp shed. At position B, the rapier head has entered the shed and reached the other end of the machine, and it catches the tip of weft yarn 3. At position C, the rapier head has retracted to the original position, laying the weft yarn inside the shed and is ready for the next cycle.

Single rapier carries the yarn in one way only, and half the movement is wasted. It is slow and occupies more space. The rapier requires high mass and rigidity. The maximum weft insertion rate of a single rapier weaving machine is 400mpm. For these reasons, single rapiers are not preferred. However, single rapier weaving machines can be used to weave weft yarns that are difficult to control.

9.1.2 DOUBLE RIGID RAPIER MACHINES

In double rigid rapier machines, two rapier heads are used, one on each side of the machine. Both rapiers start from both sides at the same time, meet at the midpoint and retreat backwards. The rapier, called a giver, takes the weft yarn from the

DOI: 10.1201/9780367853686-10

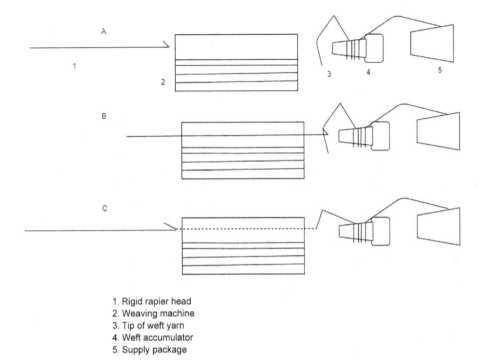

1. Rigid rapier head
2. Weaving machine
3. Tip of weft yarn
4. Weft accumulator
5. Supply package

FIGURE 9.1   Weft insertion cycle of single rapier machines.

accumulator, carries up to the centre and transfers to the other rapier called the taker. The taker rapier retracts and brings the weft yarn to the other side. Because both rapiers extend outside the loom during withdrawal, the space requirement for double rigid rapier machines is high. Figure 9.2a illustrates a double rigid rapier machine. In the figure, mark A shows the width of the loom required for a rigid rapier.

9.1.3   DOUBLE FLEXIBLE RAPIER MACHINES

Since the double rigid rapiers occupy more space the rapier rigid rods are replaced by flexible metal or plastic tapes or bands which can be wound on a drum. These machines are called flexible rapiers machines. This saves space compared to rigid rapiers. Figure 9.2b shows a double flexible rapier machine. In the figure, mark B shows the width of the loom required for flexible rapier, which is less than that required for rigid rapiers.

9.2   METHODS OF DOUBLE RAPIER WEFT INSERTION

There are two types of weft insertion in double rapier machines: tip transfer (Dewas) and loop transfer (Gabler) systems.

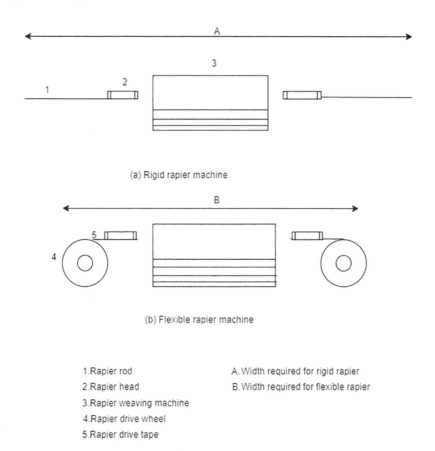

(a) Rigid rapier machine

(b) Flexible rapier machine

1.Rapier rod   A.Width required for rigid rapier
2.Rapier head   B.Width required for flexible rapier
3.Rapier weaving machine
4.Rapier drive wheel
5.Rapier drive tape

FIGURE 9.2   Double rapier weaving machines.

9.2.1   TIP TRANSFER OR DEWAS SYSTEM

In the Dewas system of weft insertion, the giver rapier grips the tip of the weft yarn, carries it to the centre of the machine and transfers it to the taker, which takes the weft to the other side of the machine during its retraction. Figure 9.3 illustrates the tip transfer or Dewas system of rapier weft insertion.

In Figure 9.3 at position A, giver rapier head 2 grips the tip of weft yarn 4 and starts moving to the centre of the machine. At the same time, taker rapier 1 also starts moving towards the centre. At position B, both the rapiers, the giver and the taker, meet at the midpoint, and the giver transfers the tip of the weft yarn to the taker and the taker grips the tip. At position C, both rapiers retreated backwards to their original positions. But the taker brings the weft yarn to its side of the machine, thereby completing the weft insertion. The giver inserts the weft up to the midpoint, and the taker inserts the weft in the remaining portion of the warp. By travelling back and forth up to the midpoint of the machine, both rapiers complete one weft insertion cycle. In modern rapier weaving machines, the tip transfer of weft insertion is mostly used due to its simplicity and speed.

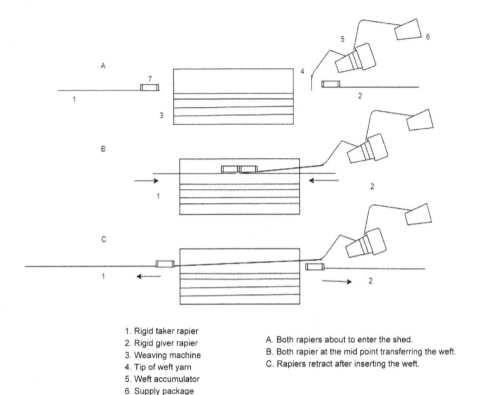

1. Rigid taker rapier
2. Rigid giver rapier
3. Weaving machine
4. Tip of weft yarn
5. Weft accumulator
6. Supply package
7. Rapier head

A. Both rapiers about to enter the shed.
B. Both rapier at the mid point transferring the weft.
C. Rapiers retract after inserting the weft.

FIGURE 9.3   Tip transfer or Dewas system of rapier weft insertion.

9.2.2   Loop Transfer or Gabler System

In the Gabler system, the giver rapier does not grip the weft; instead, it extends the weft yarn in the form of "U"-shape (loop) to the centre of the machine. The tip of the weft yarn is still held by the stopper. At the midpoint, the giver transfers the loop to the taker, and at the same time, the stopper releases the tip of the yarn. When the taker retracts, the loop of yarn straightens itself, and no further yarn is released by the accumulator. This makes straightening of the loop without slag. Figure 9.4 shows the Gabler system of weft insertion.

9.3   RAPIER HEADS

In modern rapier weaving machines, rapier heads are called grippers. Figure 9.5 shows the left-hand and right-hand rapier heads of a dornier loom.

In Figure 9.5, lever 1 is pivoted at point 2 in both the right-hand and left-hand rapier heads. The lever carries yarn clamp 3 at one end, and at the other end, it carries protrusion 4. Between the pivot and protrusion, a hardened plate 7 is secured. Under lever 1, a set of leaf springs 5 is fastened, and it applies a clamping force to the

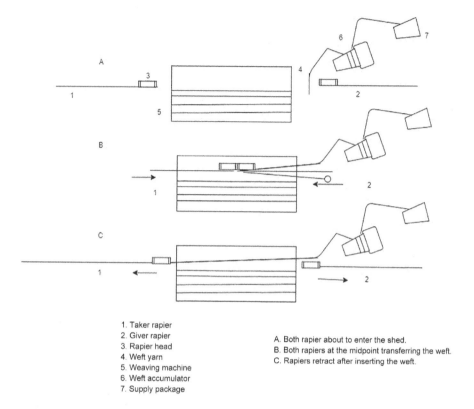

1. Taker rapier
2. Giver rapier
3. Rapier head
4. Weft yarn
5. Weaving machine
6. Weft accumulator
7. Supply package

A. Both rapier about to enter the shed.
B. Both rapiers at the midpoint transferring the weft.
C. Rapiers retract after inserting the weft.

FIGURE 9.4   Loop transfer or Gabler system of rapier weft insertion.

yarn clamps. Both the rapier heads are fastened to tape 8. The yarn clamp plates are made of different materials according to the requirement.

During operation, the weft yarn is presented to the left-hand rapier clamp, and the yarn is seized. When the weft yarn is held by the clamp, it is cut off by a cutter between the selvedge and the yarn clamp. Then weft is inserted in the warp shed.

9.4   RAPIER DRIVES

The to-and-fro movement of the rapier heads is derived from the main shaft either by a linkage mechanism or by a cam. Different makers have different designs for rigid rapier drives and flexible rapier drives. Figure 9.6 shows a flexible rapier drive system. The continuous rotary motion of main shaft 2 is converted into an oscillating motion of quadrant 4 by cam assembly 1 and connecting arm 3. The oscillating motion of the quadrant is transferred to rapier drive wheel 6 by pinion 5. When the quadrant oscillates, it makes rapier tape 7 move forwards and backwards. The gearing of the rapier tape with the drive wheel is achieved by engaging the perforations of the tape with the sprockets of the driving wheel. The rapier movement can be adjusted by adjusting the distance between fulcrum 10 and rapier adjustment point 9.

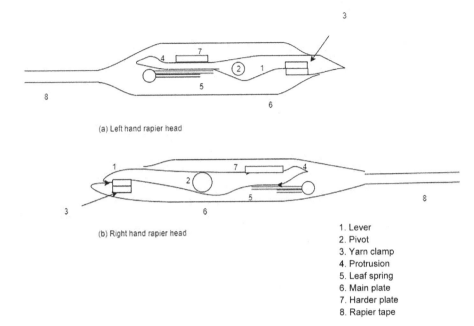

(a) Left hand rapier head

(b) Right hand rapier head

1. Lever
2. Pivot
3. Yarn clamp
4. Protrusion
5. Leaf spring
6. Main plate
7. Harder plate
8. Rapier tape

FIGURE 9.5 Left-hand and right-hand rapier heads.

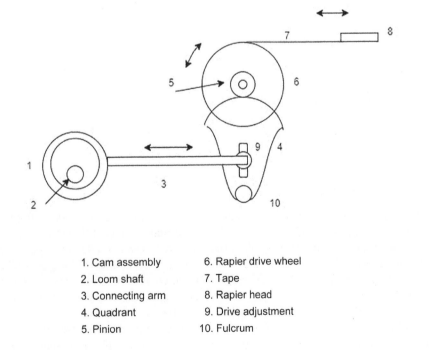

1. Cam assembly	6. Rapier drive wheel
2. Loom shaft	7. Tape
3. Connecting arm	8. Rapier head
4. Quadrant	9. Drive adjustment
5. Pinion	10. Fulcrum

FIGURE 9.6 Flexible rapier drive.

The principle of operation of the dornier rigid rapier drive is shown in Figure 9.7. By means of a pair of cams, rocking shaft 1 is rocked in its centre. This rocking is transmitted to rocking arm 4 via arm 2 and slide 3. Quadrant 5 is fastened to rocking arm 4. The quadrant is engaged with pinion 6, which, in turn, is engaged with bevel gear 10. The bevel gear transmits its movement to rapier drive spur wheel 7, which, in turn, moves rapier rod 8 forwards and backs. The relative position of the slide and the rocking arm will adjust the stroke of the rapiers.

9.5   RAPIER MOTION

The rapier heads in the rapier weaving machines are accelerated twice and decelerated twice for every pick cycle. Starting from rest at 0° (beat-up), the rapier reaches a speed of about 15m/sec at 90° and slows down to stop immediately at transfer point at 180°. In modern machines, the maximum rapier velocity is as high as 60m/sec. The rapiers have to be accelerated up to a quarter of the reed space and decelerated

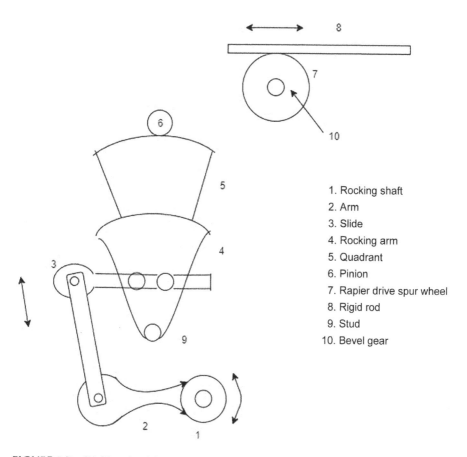

1. Rocking shaft
2. Arm
3. Slide
4. Rocking arm
5. Quadrant
6. Pinion
7. Rapier drive spur wheel
8. Rigid rod
9. Stud
10. Bevel gear

FIGURE 9.7   Rigid rapier drive.

up to the middle. Figure 9.8 shows the rapier motion with respect to machine angle of rotation for normal rapier weaving machines.

In Figure 9.8, both left- and right-hand rapiers start at 0° (beat-up position) At point 2, the weft yarn is presented to the left-hand rapier. At point 3, the left-hand rapier seizes the weft yarn. At point 4, the weft yarn between the rapier head and the selvedge is cut by the cutter. At point 5, which is about 90° of machine rotation, the left-hand rapier reaches maximum velocity. At point 6, which is at 180° of machine rotation, the weft is transferred. At point 7, the right-hand rapier reaches maximum velocity. At point 8, weft yarn is released by the right-hand rapier, and the rapier movement stops. This makes one complete weft insertion cycle of rapiers.

9.6 SALIENT FEATURES OF RAPIER WEAVING MACHINES

1. Double flexible rapier machines are the standard machines, and they can be used to weave a wide range of fabrics with natural and synthetic fibre yarn.
2. In rapier weaving machines, the sley rocks on a shaft and complementary cams, with different sley dwells are used depending on the width of the loom.

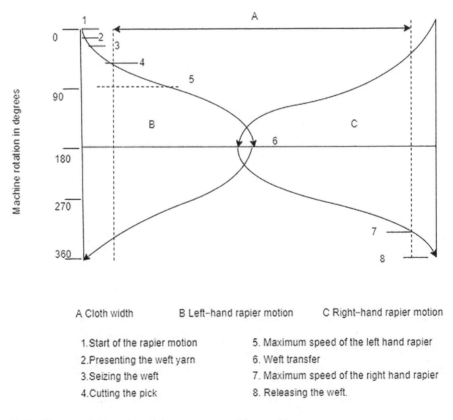

A Cloth width   B Left–hand rapier motion   C Right–hand rapier motion

1. Start of the rapier motion
2. Presenting the weft yarn
3. Seizing the weft
4. Cutting the pick

5. Maximum speed of the left hand rapier
6. Weft transfer
7. Maximum speed of the right hand rapier
8. Releasing the weft.

FIGURE 9.8 Rigid motion with respect to machine position.

3. Lightweight all-purpose grippers with few moving parts can be used for a wide range of yarns from 5tex to 1000tex.
4. Electronically controlled colour selectors can be used for selecting eight colours in weft.
5. Tuck-in, leno or fused selvedges can be formed on rapier machines depending on the requirement. The selvedge waste of weft yarn at both sides will be from 7cm to 14cm.
6. Rapier machines can be fitted with intermediate tucking units. This allows weaving several fabrics simultaneously.
7. Positive cam or positive dobby are mostly used for shedding.
8. Fabric take-up is controlled by a servomotor.
9. The weft insertion rate is up to 1100m/minute, and the width of the machine can be 220cm. The machine runs at 220ppm.

10 Air-Jet Weaving

Air-jet weaving is a shuttleless weaving technique in which compressed air is used for weft insertion as a medium. Compressed air is blown through a nozzle along with weft yarn. Due to the frictional force between air and the yarn, compressed air jet carries the weft along with it to the other side of the machine through the shed. It has an extremely high weft insertion rate. Out of all shuttleless weft insertion techniques, air-jet weft insertion is the simplest. Relative to shuttle, projectile and rapier, the mass of insertion medium is very small in air-jet weaving.

10.1 PRINCIPLE OF AIR-JET WEAVING

The important components in air-jet weaving are tandem / auxiliary and main nozzles, stopper or weft brake systems, yarn feeders and air guides, such as a confusor or profile reed. Figure 10.1 shows the schematic of air-jet weft insertion.

Weft from supply package 1 is drawn by measuring drum 2. The amount of weft yarn for each pick is measured. When compressed air 7 is blown through tandem nozzle 5 and main nozzle 6, stopper 4 opens and releases the weft yarn. Airstream 8 carries the weft to the other side of the machine. Only a predetermined length of yarn for a pick will be released by the stopper so that extra weft yarn will not be protruding outside the selvedge. This completes one cycle of pick insertion. The tandem nozzle/auxiliary nozzle airstream pulls the yarn from the measuring drum, and the main nozzle airstream gives the necessary initial acceleration to the weft yarn. The relay nozzles augment the main nozzle in carrying the weft across the warp shed. Figure 10.1a shows a main nozzle of an advanced air-jet weaving machine fitted in the sley, and Figure 10.1b shows an auxiliary nozzle of an advanced air-jet weaving machine fitted in the frame.

10.2 AIR-JET WEFT INSERTION CONFIGURATIONS

Mainly three weft insertion configurations are used in air-jet weaving machines:

1. Single nozzle with confusor guides and suction
2. Multiple nozzles with guides
3. Multiple nozzles with profile reed

10.2.1 SINGLE NOZZLE WITH CONFUSOR GUIDES AND SUCTION

In this system, a single nozzle is used to insert the weft. The air speed falls rapidly beyond a short distance from the nozzle due to the expansion of the airstream in a parabolic form. The acceleration of the yarn also falls more rapidly. To overcome this effect, a multi-ring constrictor known as a confusor is mounted on the sley over the entire width of the machine. The confusor forms an orifice

DOI: 10.1201/9780367853686-11

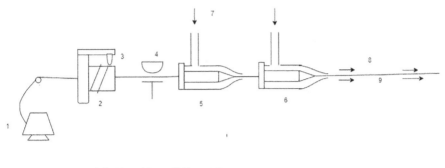

1. Supply package   6. Mian nozzle
2. Measuing drum   7. Compressed air
3. Clamp   8. Air stream
4. Stopper   9. Weft yarn
5. Tandem nozzle

FIGURE 10.1   Schematic of air-jet weft insertion.

FIGURE 10.1(A)   Main nozzle of an advanced air-jet weaving machine fitted in the sley.

and guides the airstream without loss of velocity. The confusor goes below the warp during beat-up. To augment the air velocity at the off end of the machine, suction is provided. Figure 10.2 shows a single nozzle and confusor guide system. Because the confusor lamellae have to be placed closely and have to get in and out of the warp shed for every pick, they place a large amount of stress on warp

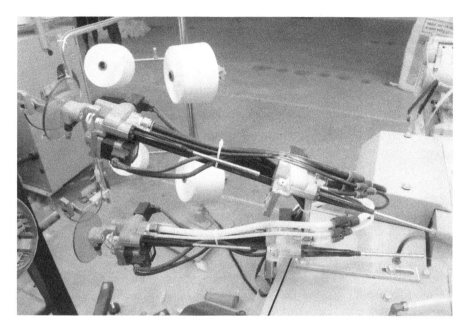

FIGURE 10.1(B)   Auxiliary nozzle of an advanced air-jet weaving machine fitted in the frame.

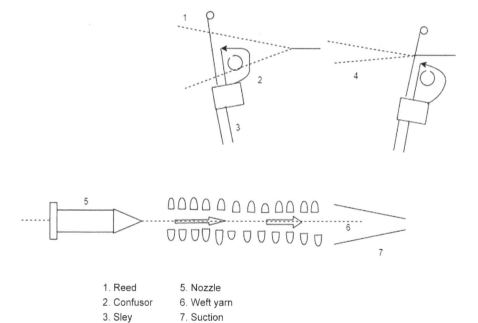

1. Reed   5. Nozzle
2. Confusor   6. Weft yarn
3. Sley   7. Suction
4. Warp shed

FIGURE 10.2   Single nozzle and confusor guide system.

yarns. Because the air velocity drops after a distance, loom width is a limitation in this system.

10.2.2 Multiple Nozzles with Guides

The disadvantage of the single nozzle system is the drop in velocity after a certain distance. In a multiple nozzle system, this has been overcome by providing auxiliary nozzles across the loom at certain intervals. Air will be injected through them in groups sequentially in the direction of the yarn movement. Suction is dispensed with in this system. Figure 10.3 shows multiple nozzles with a guide system.

10.2.3 Multiple Nozzles with a Profile Reed

In this system, the air guides are built in the reed itself as an integral part, and this type of reed is called a profile reed. The profile reed system eliminates the entrance and exit of the confusor air guides in and out of the shed for every pick. The auxiliary nozzles, also called relay nozzles, are fixed across the machine on the sley. Figure 10.4 shows multiple nozzles with a profile reed system. Figure 10.4a shows a profile reed and relay nozzles.

10.3 TIMING DIAGRAM

The typical sequence of different operations of an air-jet weaving machine with multi-nozzles and profile reed is given in Figure 10.5. The main nozzle is on at point 1, around 45°, and first relay nozzle group is opened at point 2 around 60°.

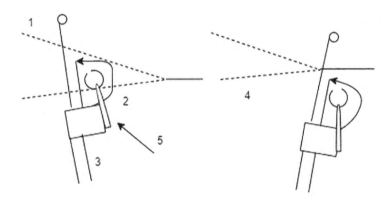

1. Reed
2. Confusor
3. Sley
4. Warp shed
5. Auxiliary nozzle

FIGURE 10.3 Multiple nozzles with a guide system.

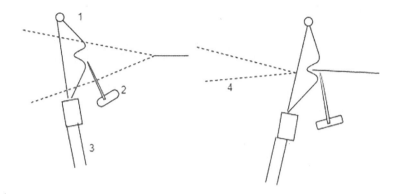

1. Profile Reed
2. Relay nozzle
3. Sley
4. Warp shed

FIGURE 10.4 Multiple nozzles with a profile reed system.

FIGURE 10.4(A) Profile reed and relay nozzles.

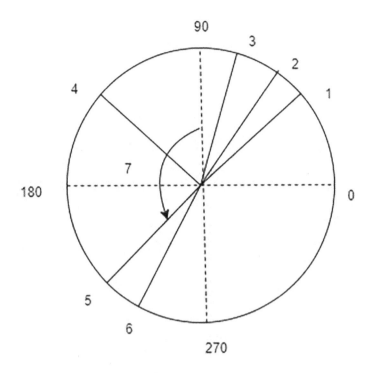

1. Main nozzle on   5. Clamp closes
2. Relay nozzle on   6. Relay nozzle off
3. Yarn release   7. yarn insertion
4. Main nozzle off

FIGURE 10.5 Timing diagram of air-jet weft insertion with multi-nozzles.

Thereafter, the yarn is released at point 3, around 80°, and weft insertion starts. At point 4, around 120°, the main nozzle is stopped. At point 5, around 230°, the yarn clamp closes, and further release of weft is stopped. The last group of relay nozzles are closed at point 6, around 260°. Thereafter, beat-up takes place. This completes one cycle of operation.

10.4 CONTROL OF AIR BLOW IN MAIN AND RELAY NOZZLES

Air-jet weaving machines are high-speed weaving machines, and precise air blow control is very essential for efficient performance. All the relay nozzles are grouped into five or six groups. The number of groups depends on the machine's width. Figure 10.6 shows the timing of air blow in main and relay nozzles.

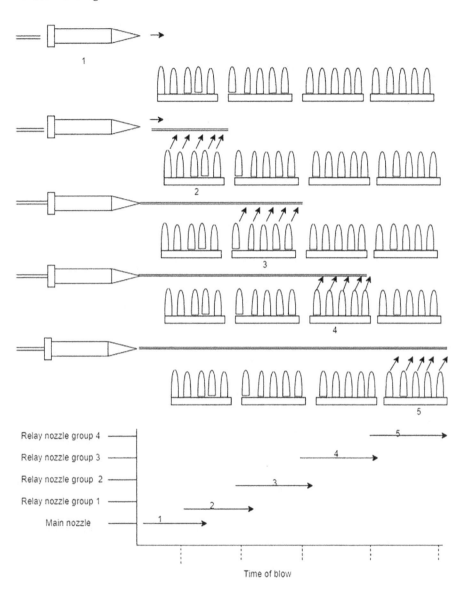

1. Main nozzle blow
2. Raley nozzle group 1 blowing
3. Raley nozzle group 2 blowing
4. Raley nozzle group 3 blowing
5. Raley nozzle group 4 blowing

FIGURE 10.6   Air blow timing in main and relay nozzles.

First, the main nozzle blow air and the weft are carried forward up to relay nozzle group 1. Then relay nozzle group 1 starts blowing at the tip of the weft yarn. This carries the weft to the next stage, and this continues up to the end. Every time, air is blown only at the tip so that the yarn is pulled throughout the insertion. This prevents buckling of yarn as well as low consumption of air.

10.5   YARN PERFORMANCE IN AIR-JET INSERTION

Since the force required to move the yarn is exclusively provided by the frictional force between the yarn and air, yarn properties such as the structure, twist, yarn diameter and fibre surface play an important role in air-jet weft insertion. Spun and coarse yarns have higher frictional coefficients. Therefore, they perform better than fine and smooth yarns. Spun yarns having a high twist, large denier, long stable and high fibril cohesion perform better in air-jet weaving. For weaving continuous filament yarns in an air jet, more air is required than for spun yarns due to their smooth surface. Larger diameter yarns require more air pressure due to their increased mass. Higher linear density increases insertion time.

Twist plays an important role in air-jet yarn insertion. Increased twist brings the fibres closer and makes the yarn more compact. This makes the yarn smoother and reduces the diameter of the yarn, which, in turn, reduces the friction between yarn and air jet, resulting in a longer insertion time. Plied yarns have a longer insertion time than single-ply yarns due to the fact that plied yarn results in a smoother yarn surface. Textured yarns increase the frictional force between the yarn and the air jet than straight filament yarn. Therefore, textured yarns take less time for insertion.

10.6   AIR-JET WEAVING MACHINES

10.6.1   SHEDDING MECHANISM

Air-jet weaving machines are equipped with negative cam shedding mechanisms for high-speed operations. Figure 10.7 shows a negative cam shedding mechanism. When non-positive cam 1 operates, it pulls heald frames 4 into a low position via cable traction 3. Once the cam operation is over, spring 5 pulls the heald frames to a high shed position. Figure 10.8 shows a negative cam tappet.

10.6.2   YARN FEEDERS

In air-jet weaving machines, specially designed drum storage feeders are used. Figure 10.9 shows a yarn feeder. The rotating yarn guide draws the weft yarn from the package and winds on the measuring drum. The electronically controlled stopper pin releases the weft yarn at the time of weft insertion. In order to minimize weft waste, it is necessary to release an exact length of weft for each pick. This is done by the stopper pin.

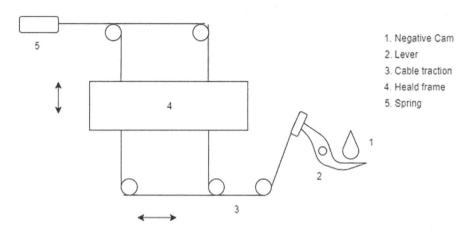

1. Negative Cam
2. Lever
3. Cable traction
4. Heald frame
5. Spring

FIGURE 10.7   Negative cam shedding mechanism.

FIGURE 10.8   Negative cam tappet.

10.6.3 SLEY MOVEMENT

The sley is driven by a crank mechanism in an air-jet weaving machine with a small arm. Figure 10.10 shows a sley movement in an air-jet weaving machine. Crank 1 oscillates sley arm 2, which, in turn, rocks sley tube 3. The rocking of the sley tube makes beat-up. Figure 10.11 shows a sley arm.

10.6.4 SELVEDGE

Leno selvedge is mainly used in air-jet weaving machines. Tucked-in selvedge is rarely used. Fused selvedge also can be formed. Mainly high-quality selvedge is formed by using full and half leno selvedge devices. Figure 10.12 shows a full-cross leno device. This device binds the weft on both sides with two leno yarns. The two

FIGURE 10.9   Yarn feeder.

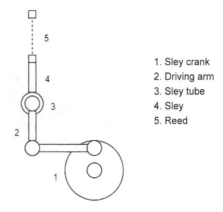

1. Sley crank
2. Driving arm
3. Sley tube
4. Sley
5. Reed

FIGURE 10.10   Sley movement.

FIGURE 10.11   Sley arm.

FIGURE 10.12   Full leno device.

spool holders with special leno yarns are rotated by a gear drive. The leno yarns unwound from the spools move up and down to produce a full-cross leno weave around the weft.

10.6.5   Air Requirement

Compressed air is used in air-jet weaving. For this purpose, separate air compressors are to be installed, and pipelines have to be arranged. The quality of air is important. The air must be free from oil and moisture. Otherwise, the nozzles will be clogged.

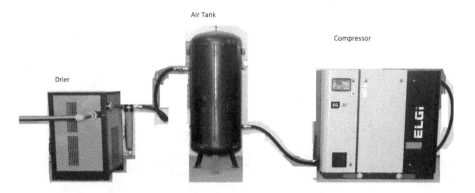

FIGURE 10.13   Compressed air plant.

Hence, oil filters and air driers have to be installed in the airline. Humidity in the air causes corrosion in the air pipes and may even cause corrosion in the machine itself. Maintaining these units involves additional cost. Figure 10.13 shows a compressed air plant system.

11 Water-Jet Weaving

Water-jet weaving machines use highly pressurized water as a medium for weft insertion. Due to the viscosity of water and its surface tension, a water jet is more cohesive and does not break up easily. It has a longer propulsive zone. Because there is no lateral force in a water jet, the weft yarn does not contort. Only a single main nozzle is used in water-jet weaving machines. Water pressure and the diameter of the jet determine the width of the weaving machine. The amount of water per pick is usually less than 2cc. A pump is used to generate the required pressure for the water jet. Figure 11.1 shows the schematic of a water-jet weft insertion. For weft feeding, a measuring drum is used to supply the weft, which supplies exact length of weft for each pick.

Weft yarn is drawn from package 1 by the measuring drum and pressure roller 3 and fed into nozzle 5. Water from sump 9 is pressurized by pump 8 and ejected as a water jet through nozzle 5. The water jet carries the weft through the shed to the other end of the machine by tractive force. The weft tends to fall as it moves from the nozzle to the other end. Hence, its travel path is arranged in an arc form. This is achieved by adjusting the nozzle holder at an angle.

The weft is cut at the selvedge by heaters and a fused selvedge is formed. The heater temperature is around 500 °C.

All yarns cannot be used in water-jet weaving machines. The yarn must be hydrophobic in nature and insensitive to water. Thermoplastic yarns, such as nylon, polyester, polypropylene and glass, are successfully woven in water-jet weaving machines. Heaters are used to dry the cloth. The wastewater after weft insertion is collected through pipelines.

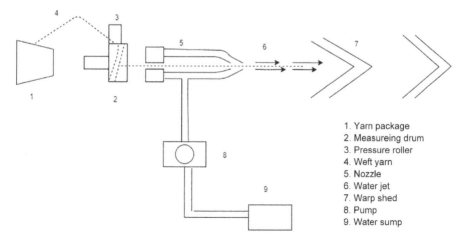

1. Yarn package
2. Measureing drum
3. Pressure roller
4. Weft yarn
5. Nozzle
6. Water jet
7. Warp shed
8. Pump
9. Water sump

FIGURE 11.1   Schematic of water-jet weft insertion.

DOI: 10.1201/9780367853686-12

Water-jet looms are the least flexible. The fabric woven must be hydrophobic, and only medium-weight fabrics can be woven successfully, even though lightweight fabric can be woven with great difficulty. Yarn wastage is more in water-jet looms. Nevertheless, water-jet looms are best suited to produce 100% filament fabrics on a mass scale. Water-jet looms use less energy and are the least noisy.

12 Multiphase Weaving

The main weaving motions of a classical weaving machine, namely shed formation, weft insertion and beat-up, take place along the full width of the machine. These weaving machines are called single-phase weaving machines. All single-phase weaving machines or looms form one shed at a time, one weft insertion at a time and one beat-up at one time. In a multiphase weaving machine, several weft insertions take place simultaneously, and several shed formations take place at a time.

There are two principles of multiphase weaving machines:

1. Weft direction shed wave principle
2. Warp direction shed wave principle

12.1 WEFT DIRECTION SHED WAVE PRINCIPLE

In weft direction shed wave principle machines, a number of sheds are formed across the entire width of the machine for the insertion of weft. These sheds look like a wave from one side to the other. The weft carrier slides into each shed. As the weft carriers enter one section of the warp, a shed is formed, and as it leaves to the next section, another shed is formed in the next section. This will occur along the full width of the machine. As a result, at any time several shuttles will insert different weft yarn. Figure 12.1 shows weft direction shed wave principle. Due to the insertion of weft by many shuttles at a time, the weft insertion rate is very high.

The weft yarn is beaten up by a rotating reed. Figure 12.2 shows a beating up by a rotary reed.

Individual reed blades 1 are assembled in rotating shaft 2. When the reed blade rotates, weft yarn 5 is caught by groove 3 and pushed towards the fell of the cloth. As the reed blade rotates further, the weft yarn slips out of the groove, but the reed blade pushes the weft to the fell of the cloth.

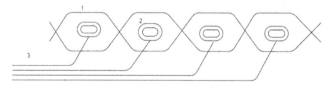

1. Warp Sheds
2. Shuttles
3. Weft yarns

FIGURE 12.1 Warp direction shed wave principle.

DOI: 10.1201/9780367853686-13

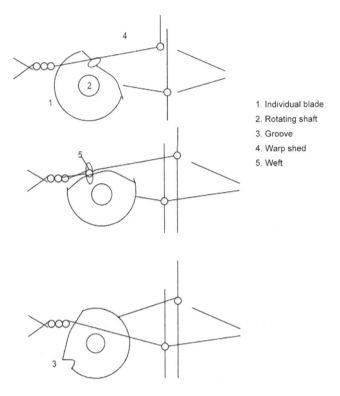

1. Individual blade
2. Rotating shaft
3. Groove
4. Warp shed
5. Weft

FIGURE 12.2   Beating-up by rotary reed.

12.2   WARP DIRECTION SHED WAVE PRINCIPLE

In these machines, several sheds are opened in the warp direction, one behind the other across the full width of the machine simultaneously. Weft yarns are inserted into each of these sheds simultaneously. Either rapier or air jet can be used for weft insertion. Figure 12.3 shows the warp direction shed wave principle

A weaving rotor is used for shed formation. Figure 12.4 shows a weaving rotor. The warp ends pass over weaving rotor 4. The sheds are formed consecutively by shed-forming elements 1 on the circumference of the weaving rotor. The motion of the rotor causes the shed forming elements to open the sheds one after the other. Weft is inserted into the warp sheds across the full width of the machine either by rapier or by air jet. The beat-up combs placed between the shed-forming elements push the weft yarn to the fell of the cloth and beat up the weft.

Due to the simultaneous weft insertions by several weft carriers, the weft insertion rates of multiphase weaving machines are very high, between 3000mpm and 5000mpm. Heavy fabrics cannot be woven in multiphase weaving machines due to the nature of beat-up in these machines.

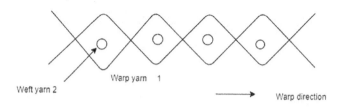

FIGURE 12.3   Warp direction shed wave principle.

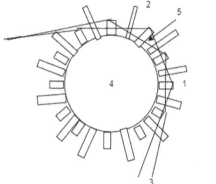

1. Shed-froming element
2. Beat-up comb
3. Warp yarn
4. Weaving rotor
5. Weft channel

FIGURE 12.4   Weaving rotor.

12.3   CIRCULAR WEAVING MACHINES

The circular weaving machine has a circular warp and continuously circulating shuttles run around the periphery of the machine in the wave or ripple shed. Circular weaving permits the use of several shuttles at the same time, and hence, productivity is high. Due to lack of flexibility, only two articles, namely sacks and tubes, can be woven in circular weaving machines.

On a circular weaving machine, each shuttle runs in its own shed. The warp is divided into segments, which forms sheds with small heald frames or wires. Heald frames are controlled by cams, and only plain and twill weaves are possible. The shuttles can be driven by either mechanical or electrical means. The weft is not beaten up just like that of single-phase weaving machines. The needle gears follow the shuttle and push the weft to the fell of the cloth.

13 Techno-Economics of Shuttleless Looms

There are mainly four types of shuttleless loom used for mass production: projectile, rapier, air-jet and water-jet looms. They differ in principle of operation and, hence, their operational cost. The manufacturing cost may be different for different looms for producing the same fabric. The techno-economics of all the types of shuttleless looms are discussed herewith.

13.1 TECHNO-ECONOMICS OF PROJECTILE LOOMS

The projectile looms have many moving parts, and their manufacturing cost is very high compared to other types of looms. The projectile picking system involves many sub-mechanisms. They are the costliest looms. The cost of a 340cm-width new projectile loom (P7300 HP) is approximately Rs. 15 million. It is highly uneconomical to produce grey sheeting in these machines when other technology machines are available at a cheaper rate with the same speed. Mainly the air-jet looms have made inroads in their territory. As per the industry sources, now a days, new projectiles looms are no longer economical for producing grey sheeting. No new machines are installed for producing ordinary regular fabrics. They are used to produce only specialty fabric such as technical textiles and denims.

But the techno-economics of second-hand projectile looms are entirely different. Second-hand machines are available at Rs. 1 million in the international market with residual life of 20 years. Many weaving factories in EU countries, Italy, Brazil and Argentina are closing their operations, and the projectile looms running in these countries are being exported to India in large numbers. The reason for a large share of imported shuttleless looms being second-hand projectile looms is the fact that the cost is almost three times less compared to new air-jet looms, but at the same time, the quality of fabric produced in these machines is equivalent to the quality of fabric produced in new air-jet looms. The preferred models of second-hand projectile shuttleless looms are Sulzer make P7100, P7150 and TW11.

13.2 TECHNO-ECONOMICS OF AIR-JET LOOMS

The air-jet looms have the least moving parts compared to other types of looms. The maintenance expenses for air-jet looms are less. As per the industry source, the cost of a 190cm Japanese brand-new air-jet loom running at 1000rpm is approximately Rs. 3.8 million. The cost of 280cm- and 340cm-weaving-width air-jet looms is Rs. 4.4 million and Rs. 5.8 million, respectively. The 280cm loom runs at 800rpm, and the 340cm loom runs at 600rpm. The industry has the flexibility in choosing the air-jet looms according to their requirements and investment plan.

DOI: 10.1201/9780367853686-14

Hence, air-jet looms are preferred by the industry for new investment. The disadvantage of air-jet looms is their power cost. However, the overall fabric manufacturing cost is less. The preferred model of air-jet loom is Tsudakoma ZAX9100. The demand for air-jet machines is higher than for projectile machines in the Indian industry.

13.3 TECHNO-ECONOMICS OF RAPIER LOOMS

Rapier weaving machines are preferred for designed fabrics due to their flexibility in coloured weft selection. They are generally not preferred for producing grey sheeting due to their higher weft yarn wastage. In Bhiwandi, one of the largest textile clusters in India, of 25,000 shuttleless looms, 90% of the shuttleless looms are Chinese rapiers. This is because of the cheaper price compared to the highly prized shuttleless looms from EU or Japan. Erode, another textile cluster in India that is famous for producing checked fabrics, of the 3200 shuttleless looms installed, most of them are rapier machines.

13.4 TECHNO-ECONOMICS OF WATER-JET LOOMS

In water-jet looms, only synthetic yarns can be used for weaving, and good quality water is required. In India, Surat meets these two criteria. Out of a total of 22,600 shuttleless looms running in Surat, about 20,000 looms are water-jet looms. Nearly 30 million metres of raw synthetic fabric and 25 million metres of processed fabric are produced in Surat daily.

13.5 PROJECT COST OF LOOMS AND MANUFACTURING COST OF FABRICS

The production of cotton and viscose rayon grey fabric is carried out in large quantities using new imported air-jet looms and second-hand imported projectile looms in South India. The use of rapier looms is restricted to producing only designed fabrics and water-jet looms are used only to produce synthetic fabrics. Therefore, the competition is only between new air-jet looms and second-hand imported projectile looms. To evaluate the performance of these looms in terms of manufacturing cost and investment required, face-to-face interviews with entrepreneurs were carried out in the Coimbatore region, where more than 5000 air-jet looms and 5000 second-hand projectile looms are running. The details of the interviews follow.

13.5.1 PROJECT COST AND MANUFACTURING COST OF FABRIC FOR NEW AIR-JET LOOMS

A new factory commissioned recently in the Palladam area of the Coimbatore region of South India, with 64 new air-jet looms of Japanese make, was chosen for calculating the project cost for installing new air-jet looms and the manufacturing cost. Based on the face-to-face interview with the entrepreneur, the capital investment required to install 64 new air-jet looms of Japanese make is given in Table 13.1, and the running expenses per month are given in Table 13.2.

The fabric produced in the air jet looms is viscose rayon grey sheeting. The warp and weft count of yarn used is 30s Ne. Ends per inch is 68, and picks per inch is 60. The width of fabric is 63 inches. Because the raw material cost and selling price of fabric vary frequently according to market fluctuations, the conversion cost offered by the fabric merchants per metre of fabric is taken for calculating the total production and the profit calculation. The details are given in Table 13.3.

Total investment (excluding land cost) for installing 64 new Japanese make air-jet loom is Rs. 22,70,04,190, and the manufacturing cost per metre of fabric is Rs. 5.32. The simple payback period is 8 years 9 months.

TABLE 13.1
Capital Investment for New Air-Jet Looms (64 Nos.)

S No	Details	Required	Cost per Unit in Rupees	Total Cost in Rupees
1	Building	2045 sq. m	9782/sq. m	2,00.04,190
2	Shuttleless looms (air-jet new) 190cm width	64	30,00,000/loom	19,20,00,000
3	Power line cables			40,00,000
4	Air compressor			50,00,000
5	Pipeline			15,00,000
6	Air drier, air tank, etc.			20,00,000
7	Warehouse machines			25,00,000
	Total investment (excluding land cost)			Rs.22,70,04,190

TABLE 13.2
Running Expenses for New Air-Jet Looms (64) per Month

S. No	Details	Units	Cost in Rs	Total cost
1	Fitter	3	50,000 per month	1,50,000
2	Gaiter	6	19,000	1,14,000
3	Shift fitter	4	25,000	1,00,000
4	Weavers @ 4 loom per weaver	960 man days	540/man day	5,18,400
5	Spares and oil	64 looms	2000/loom/month	1,28,000
6	Manager	1	60,000/month	60,000
7	supervisor	2	30,000/month	60,000
8	Shift-in-charge	2	15,000/month	30,000
9	Electricity charges		Per month	22,40,000
10	Total running expenses			34,00,400
11	Interest charges per month @9 % per annum for the total investment of Rs 22,70,04,190 (Land cost not included)			17,02,531
12	Total expenses per month including interest			51,02,931

TABLE 13.3
Total Production and Profit Calculation (new air-jet looms)

S. No	Details	Units
1	Width of loom	190cm
2	Speed of looms	1000rpm
3	Number of looms	64
4	No of fabrics per loom	1
5	Production/day/loom @90 % efficiency	500m
6	Production per month (30 days) for 64 looms (500 × 64 × 30)	9,60,000m
7	Conversion cost offed per metre of fabric	Rs 7.54
8	Total income/month	Rs.72,38,400
	(9,60,000m × Rs 7.54)	
9	Net profit per month (Total income – Total expenses)	21,35,469
10	Simple payback period (Total investment divided by profit per year)	8 years 9 months
11	Manufacturing cost per metre of fabric	Rs 5.32/m
	(Total expenses divided by Fabric produced per month)	

13.5.2 Project Cost and Manufacturing Cost of Fabric for Second-Hand Projectile Looms

A factory commissioned recently in the Tirupur area of the Coimbatore region of South India, with 20 second-hand imported projectile looms, was chosen for calculation of project cost for installing imported second-hand projectile looms and manufacturing cost calculation. Based on the face-to-face interview with the entrepreneur the capital investment required to install 20 second-hand imported projectile looms is given in Table 13.4. and the running expenses per month are given in Table 13.5.

The fabric produced in the second-hand projectile looms is viscose rayon grey sheeting. The warp and weft count of yarn used is 30s Ne. Ends per inch is 68, and picks per inch is 60. The width of fabric is 63 inches. Since the raw material cost and selling price of fabric varies frequently according to market fluctuations, a

TABLE 13.4
Capital Investment for Second-Hand Projectile Looms (20 Nos.)

S No	Details	Required	Cost per Unit in Rs	Total Cost in Rs
1	Building	930 sq. m	7520/sq. m	69,93,600
2	Second-hand projectile looms (390cm width)	20	10,00,000	2,00,00,000
3	Repair and installation expenditure for looms	20	1,00,000	20,00,000
4	Power line cables			5,00,000
5	Warehouse machines			2,00,000
6	Generator	1	7,00,000	7,00,000
	Total investment (Excluding land)			3,03,93,600

conversion cost offered by the fabric merchants per metre of fabric is used for calculation purposes. The total production and profit calculation is given in Table 13.6.

Total investment (excluding land cost) for installing 20 second-hand Sulzer projectile looms (TW 11) is Rs. 3,03,93,600, and the manufacturing cost per metre of fabric is Rs 5.75. The simple payback period is 12 years.

Even though the investment is high, due to lower manufacturing cost per meter of fabric in a new air-jet loom, than second-hand projectile looms, there is a tendency in the industry to go for new air-jet looms.

TABLE 13.5
Running Expenses for Second-Hand Projectile Looms per Month

S. No	Details	Units	Cost in Rs	Total cost
1	Fitter	1	40,000 per month	40,000
2	Fitter helper	2	20,000	40,000
3	Sweeper	1	10,000	10,000
4	Weavers @ 4 loom per weaver	300 man days	500/man day	1,50,000
5.	Driver	1	17000	17,000
5	Spares and oil	20 looms	4000/loom/month	80,000
6	Manager	1	19,000/month	19,000
7	Folding man	3	15000/month	45000
8	Electricity charges		Per month	2,00,000
9	Total expenditure			6,01,000
10	Interest charges per month @ 9% per annum for the total investment of Rs 3,03,93,600 (land cost not included)			2,27,952
11	Total expenses per month including interest			8,28,952

TABLE 13.6
Total Production and Profit Calculation (projectile looms)

S. No	Details	Units
1	Make	Sulzer projectile TW 11
2	Loom width	390cm
3	Speed	200rpm
4	No of fabrics per loom	2
5	No of looms	20
6	Production/day/loom @ 90% efficiency (2 × 120m)	240m
7	Production per month (30days) for 20 looms (240m × 20 × 30)	1,44,000m
8	Conversion cost offed per metre of fabric	Rs 7.20
9	Total income/month (1,44,000m × Rs 7.20)	Rs.10,36,800
10	Net profit per month (Total income – Total expenses)	2,07,848
11	Simple payback period (Total investment divided by profit per year)	12 years 2 months
12	Manufacturing cost per metre of fabric (Total expenses per month divided by fabric produced per month)	Rs 5.75/m

14 Global Weaving Industry

14.1 GLOBAL PRODUCTION OF TEXTILES

The global textile industry was estimated at around US$920 billion in 2018, and it is projected to reach approximately US$1230 billion by 2024, representing a steady compound annual growth rate of nearly 5% during the forecast period. On the basis of raw material, cotton dominated the textile market, with a value of US$378.6 billion in 2019 owing to its properties, such as high absorbency and strength. In terms of volume, polyester accounted for a share of 28.0% in 2019 and is expected to register a significant growth rate over the forecast period owing to its properties, such as excellent shrink resistance and high strength. Fashion and clothing emerged as the largest application segment, with a value of US$712.3 billion in 2019 owing to the rapid rise in consumer spending on apparel and clothing.

China is currently the world's largest textile producer and exporter of both raw textiles and garments, accounting for over half of the global textile output every year. Low cost and vast labour availability, reduced commercial barriers and strong material supply are few of the competitive advantages the country offers for the textile manufacturing industry. Regarding material supply, China produced about 79 billion metres of cloth in 2017 alone and 5.99 million metric tons of cotton in 2017/2018.

India is another one of the world's largest textile-producing countries and largest textile exporter. It holds an export value of more than US$30 billion a year. India is responsible for more than 6.9% of the world's total textile production, and its textile industry is valued at approximately US$150 billion. India is the second-largest textile producer in the world in terms of production volume. The Indian textiles industry, currently estimated at around US$150 billion, is expected to reach US$250 billion by 2019. According to the latest report from IBEF, India's textiles industry contributed 7% of the industry output (in value terms) of India in 2018–2019. It contributed two per cent to the gross domestic product (GDP) of India and employs more than 45 million people in 2018–2019. The sector contributed 15% to the export earnings of India in 2018–2019.

The United States is the leading producer and exporter of raw cotton, while also being the top importer of raw textiles and garments. The US ranked on the third place in the list of world's largest textile producing countries. According to the recent report from the US National Council of Textile Organizations (NCTO), the total value of US man-made fibre and filament, textile, and apparel shipments totalled an estimated US$76.8 billion in 2018; this is an uptick from the US$73 billion in output in 2017. Due to its productivity, flexibility and innovation, the US continues to be one of the largest textile producers in the world.

Over the recent years, many other developing countries are also gaining immense growth in their textile industries as their investment into the textile or garment industry increases. Countries such as Pakistan, Indonesia, Thailand and Sri Lanka, as well as

DOI: 10.1201/9780367853686-15

TABLE 14.1
Total Textile Output (top 10 countries)

Rank	Textile Output in Global Share (%) 2019		Textile Export Value (US$ billion) 2018	
	Country	Textile Output in the Global Share	Country	Export Value
1	China	52.2%	China	118.5
2	India	6.9%	EU	74.0
3	United States	5.3%	India	18.1
4	Pakistan	3.6%	US	13.8
5	Brazil	2.4%	Turkey	11.9
6	Indonesia	2.4%	South Korea	9.8
7	Turkey	1.9%	Taiwan	9.2
8	South Korea	1.8%	Vietnam	8.3
9	Thailand	1.1%	Pakistan	8.0
10	Mexico	0.9%	Hong Kong	7.4

Source: WTO—World Trade Organisation.

a number of South American countries, have also seen considerable growth in their textile output in recent years. As China moves towards a service-based economy and labour prices continue to rise, it is expected that more and more international buyers and investors will shift their business focus to these fast-evolving markets in the near future.

The total textile production of the top 10 countries in terms of percentage in global share and export value is given in Table 14.1.

China, ranked first in textile production in the world, contributes 52.2% of global textile output. China produced about 79 billion metres of cloth in 2017. India is the second-largest textile producer in the world in terms of production volume, with a 6.9% global share. The US ranked third in the list of the world's largest textile-producing countries, with a 5.3% global share. Pakistan, with 3.6% global share, stands fourth; Brazil, with 2.4%, stands fifth; and Indonesia, with 2.4%, stands sixth in global contribution of textiles.

14.2 FABRIC AND MANUFACTURING COST AT GLOBAL LEVEL

The manufacturing cost of textiles depends on many factors, such as raw material cost, labour cost, cost of investment, power cost, land and building cost and machinery cost. These things vary from country to country. For example in developing countries such as India and Pakistan, the cost of labour is cheaper than in other countries. Similarly, India is a cotton-producing country, and the raw material cost is lower. Due to the global economy, any country can import textile goods from anywhere in the world. Hence, the cheapest-producing country will have the advantage of more demand. It becomes necessary to understand the cost of production

TABLE 14.2

Manufacturing Cost and Total Woven Fabric Cost (ring yarn)—2008

Country Cost Element	Brazil	China	Egypt	India	Italy	Korea	Turkey	US
Total manufacturing cost (USD per meter of fabric)	0.236	0.214	0.186	0.265	0.579	0.263	0.308	0.346
Index (Italy: 100)	(41)	(37)	(32)	(46)	(100)	(45)	(53)	(60)
Total fabric cost (US D per meter of fabric)	0.858	0.848	0.842	0.827	1.386	0.877	0.866	0.959
Index (Italy: 100)	(62)	(61)	(61)	(60)	(100)	(63)	(62)	(69)

Source: Compendium of International Textile Statistics—2009–10.

of fabrics in major textile-producing countries. The manufacturing costs and total fabric costs including raw material cost per metre of fabric in US dollars for major fabric-producing countries is given in Table 14.2.

The index for manufacturing cost and total cost by taking Italy as 100 is calculated for every country. The manufacturing cost index is lowest for Egypt, with 32, and the second lowest being China, with 37. The index for total fabric cost is lowest for India, with 60, and the second lowest being Egypt and China, with 61.

The International Textile Manufacturers Federation has published International Production Cost Comparison (IPCC) for 2018. The IPCC is designed to trace the implications of the growing capital intensity in the primary textile industry. The IPCC describes manufacturing and total costs of yarn/fabric broken down into various cost elements at different stages of the textile value chain. The report covers Pakistan, Bangladesh, Brazil, China, Egypt, India, Indonesia, Italy, Korea, Turkey, the US and Vietnam. The report enlists the relative importance of the cost elements and their respective influence on the total costs.

14.2.1 Manufacturing Cost of Spinning Ring Yarn

The report shows that in spinning ring yarn (30s Ne) countries with high manufacturing cost also have a higher share of labour cost. Table 14.3 shows manufacturing cost in USD/kg of yarn for 30s Ne and share of labour cost and power cost in manufacturing cost in percentage

14.2.2 Manufacturing Cost for Fabric

The share of labour cost is higher in manufacturing cost for fabric for Italy, Korea and the US, but the power cost for these countries shows less share than the average power cost. Table 14.4 shows the share of power cost to manufacturing cost for fabric (2018) for different countries.

TABLE 14.3

Manufacturing Cost in USD/kg of Yarn for 30s Ne and Share of Labour Cost and Power Cost (2018)

S No	Country	Manufacturing Cost in USD/kg of yarn for 30s Ne	Share of Labour Cost in Manufacturing Cost in Percentages	Share of Power Cost in Manufacturing Cost in Percentage
1	Italy	$ 2.35	33	21
2	Korea	$ 1.60	27	21
3	US	$ 1.54	31	10

TABLE 14.4

Share of Power Cost to Manufacturing Cost for Fabric (2018)

S. No	Country	Share of Power Cost to Manufacturing Cost for Fabric
1	Italy	17%
2	Korea	17%
3	US	9%
4	Indonesia	30%–40%
5	India	30%–40%
6	Brazil	30%–40 %
7	Pakistan	30%–40 %
8	China	30%–40%
9	Average	26%

The share of power cost to total manufacturing cost for Indonesia, India, Brazil, Pakistan and China is between 30% and 40%. The average power cost is 26% in the manufacturing cost of fabric.

14.2.3 SHARE OF LABOUR COST AND POWER COST IN MANUFACTURING COST FOR KNITTED FABRIC

The share of labour cost and power cost are equal in manufacturing cost for knitted ring yarn fabric for China, Brazil and Egypt, with 20%, 14% and 6%, respectively. Table 14.5 shows the share of labour cost and power cost in manufacturing cost in knitted fabrics.

The labour cost is more by many times than power cost in US, Italy, and Korea while it is many times less than the power cost in India, Bangladesh and Pakistan.

14.2.4 SHARE OF LABOUR COST AND POWER COST IN MANUFACTURING COST FOR FINISHED WOVEN FABRIC

The share of labour cost and power cost in manufacturing cost for finished woven fabric is given in Table 14.6.

TABLE 14.5

Share of Labour Cost and Power Cost in Manufacturing Cost in Knitted Fabric

S No	Country	Share of Labour Cost in Manufacturing Cost	Share of Power Cost in Manufacturing Cost	Labour Cost to Power Cost Ratio
1	China	20%	20%	–
2	Brazil,	14%	14%	–
3	Egypt	6%	6%	
4	US	–	–	14 times more expensive
5	Italy	–	–	8 times more expensive
6	Korea	–	–	4 times more expensive
7	India	–	–	3 times less expensive
8	Bangladesh	–	–	3.5 times less expensive
9	Pakistan	–	–	4 times less expensive

TABLE 14.6

Share of Labour Cost and Power Cost in Manufacturing Cost for Finishing Fabric

Country	Labour Cost	Power Cost
Average percentage to manufacturing cost	13%	14%
Bangladesh	1 cent/m (USD)	–
Italy	15 cents/m (USD)	–
Egypt, Vietnam	–	3 cents/m (USD)
Brazil, Italy	–	7 cents/m (USD)

On average, the shares of labour and power to total manufacturing costs for finishing are 13% and 14%, respectively. Strong geographical discrepancies nevertheless exist, especially with respect to labour costs with a spread of 14 cent/m. This reflects the difference in labour costs between Bangladesh (1 cent/m [USD]) and Italy (15 cents/m [USD]). The spread in power cost is measured at 4 cents/m, which corresponds to the difference between the cost of power in Egypt or Vietnam (3 cents/m [USD]) and in Brazil or Italy (7 cents/m [USD]).

14.3 TEXTILE SCENARIO IN INDIA

The manufacturing cost per metre of fabric in India is US$0.265. Even though the manufacturing cost is higher in India than in many countries like Brazil, China, Egypt and Korea, due to the lower raw material cost, the total fabric cost per metre is lowest in India at US$0.827. India plays an important role in textile production by being the cheapest producer of textile fabrics with larger quantity at a global level. Therefore analyzing the techno-economics of shuttleless looms in the Indian context is important.

As per the report of the working group on Textile & Jute Industry for the XII Five Year Plan (2012–2017) of Government of India, there were approximately 5.18 lakh power loom units with 22.92 lakh power looms as on 31.03.2011 in India. The technology level of this sector varies from obsolete plain looms to high tech shuttleless looms. There are approximately 1,05,000 shuttleless looms in this sector. It is estimated that more than 75% of the shuttle looms are obsolete and outdated, with a vintage of more than 15 years. The requirement of auto/shuttleless looms for the XII Five Year Plan (2012–2017) for projected cloth production has also been calculated. An incremental 2,34,563 auto/shuttleless looms will be required to produce an additional projected cloth production of 31.39 billion square metres of fabric. If it is assumed that half of the incremental production of fabric would be by auto looms and half would be by shuttleless looms, then the requirement of shuttleless looms would be 1,17,282 during XII Five Year Plan (2012–2017).

14.4   TECHNOLOGICAL LEVEL OF WEAVING INDUSTRY IN INDIA

Indian weaving industry is modernizing itself at a rapid pace with the installation of shuttleless looms for mass production. The indigenous shuttleless loom production is very low with inferior quality. Due to this fact, a majority of the shuttleless looms installed in India are imported. The details of shuttleless looms imported from other countries to India for a period of 5 years from 2007 to 2012 is given in Table 14.7.

Out of 36,570 shuttleless looms imported, the share of rapier looms was highest at 16,186, followed by air-jet looms at 8068. Preference for rapier looms is comparatively very high compared to air-jet, projectile and water-jet looms during this period. But it has been observed that the majority of the shuttleless looms imported are second-hand

TABLE 14.7

Import of Shuttleless Looms to India

Year / Loom Type	No. of Shuttleless Looms Imported Year-Wise					Total
	2007–2008	2008–2009	2009–2010	2010–2011	2011–2012	
Air jet	1452	660	1592	2827	1537	8068
Rapier	2520	1686	4251	3873	3856	16186
Water jet	536	246	1474	1497	1577	5330
Projectile	734	407	2060	2161	1624	6986
Total	5242	2999	9377	10358	8594	36570

Source: Textile Machinery Manufacture Association.

TABLE 14.8
Import of Second-Hand Shuttleless Looms

Year \ Loom Type	No. of Second-Hand Shuttleless Loom Imported Year-Wise					Total
	2007–2008	2008–2009	2009–2010	2010–2011	2011–2012	
Air jet	339	153	900	1062	1311	3765
Rapier	1801	548	2644	3719	3637	12,349
Water jet	536	246	1474	1497	447	4200
Projectile	734	407	2060	2161	1624	6986
Total	3410	1354	7078	8439	7019	27,300

Source: Textile Machinery Manufacture Association.

due to the fact that the cost of new shuttleless looms of EU and Japan make are very high. This is beyond the reach of Indian entrepreneurs who are operating at relatively small scale. Table 14.8 shows the details of imported second-hand shuttleless looms

Out of 36,570 shuttleless looms imported, 27,300 were second-hand machines. The reason for this is that the second-hand looms, having a residual life of 10 to 20 years, are working smoothly, with 80% to 90% efficiency, and are of great help for modernizing the power loom industry. The difference between the prices of new and second-hand shuttleless looms are more than three times, especially in the case of air-jet and projectile looms.

14.5 CLUSTER-WISE STATUS OF WEAVING INDUSTRY IN INDIA

India is a vast country with different regions with different resources. Therefore, the development of textile industry also is concentrated in certain regions. Similarly, the modernization of the weaving industry and the machinery requirement also is different in different regions. The status of the weaving industry in the important regions is given in Table 14.9.

TABLE 14.9
Cluster-Wise Power Looms and Shuttleless Looms Installed (up to 2012)

Type of Loom \ Name of Cluster	Bhiwandi	Ichalkaranji	Erode	Surat	Bhilwara	Malegaon
Power looms	6, 00,000	No data	89,000	6,00,000	13,000	6,00,000
Total shuttleless looms	25,000	7500	3200	22,600	11,700	–
Rapier	22500	–	3200	2000	115	
Projectile	–	6375	–	600	3980	
Air jet	–	–	–	–	7490	
Water jet	–	–	–	20,000	115	

In the Bhiwandi cluster, there are more than 6 lakh power looms, of which about 25,000 are shuttleless looms, and most of them are rapier looms. In the Ichalkaranji cluster, out of 7500 shuttleless looms, 85% are second-hand imported projectile looms, 13% are brand new imported and only 2% are of Indian make. A majority of the units in the cluster are doing job work. For job work, second-hand machines are good. It would not be economical to install new machines for carrying out job work.

In the Erode cluster, there are about 89,000 power looms, of which only 3200 are shuttleless looms. Most of the shuttleless looms are second-hand rapier looms purchased mainly from Italy, Germany and Turkey at an average cost of Rs. 6.5 to 9.5 lakhs. Water-jet shuttleless looms are not at all used in this region due to a scarcity of water.

The Surat cluster has 6 lakh power looms, and only 22,600 are shuttleless looms. Out of a total of 22,600 shuttleless looms, about 20,000 looms are water jet, 2000 are rapier and there are only about 600 projectile looms. Since the major product of the cluster is synthetic textiles, hence, the cluster has a natural inclination towards water-jet machines.

In Bhilwara, there are about 13,000 power looms in the cluster and 11,700 are shuttleless looms. Out of these, 7490 are air-jet looms, 3980 are projectile looms and 230 are rapier and water-jet looms. The industry believes that the future of textile machinery is with air-jet looms as they have higher speeds (rpms), and the current job rate on air-jet looms is almost double that of projectile machines. The speed of projectile machines is also less than that of air-jet looms. The performance of second-hand looms is comparable with that of the brand-new looms, but the cost of the looms is a major factor behind the popularity of second-hand looms in the cluster.

Malegaon is a major textile-producing centre which produces fabric of a relatively lower quality, with more than 3 lakh power looms producing about 10 million meters of cloth annually. No shuttleless looms are installed so far.

14.6   FUTURE TRENDS OF GLOBAL TEXTILE INDUSTRY

One of the major global textile industry trends that has been witnessed is the rising demand for nonwoven fabrics, mainly driven by the downstream industries, such as the personal care, packaging, automotive and constructions sectors. Nonwoven fabrics are raw materials for producing hygiene and personal care products, such as baby diapers, sanitary napkins and adult incontinence. In the construction industry, nonwoven fabrics are used in road building in the form of geotextiles to increase the durability of roads; the automobile industry also manufactures a large number of exterior and interior parts using nonwoven fabrics. With the rising demand caused by the extensive applications in the downstream industries, the nonwoven segment is expected to witness a faster growth rate among the global textile industry over the next few years.

The rising application of smart textile products, owing to the miniaturization of electronic components and the use of conductive materials, is expected to drive product demand. Technological innovation in terms of the development of new upholstery products derived from coated fabrics and spider silk is expected to open new market opportunities in the near future.

REFERENCES

1. Mordor Intelligence (Grand View Research)
2. World Trade Organization.
3. Study Report on Impact of Imported Second Hand Shuttle-less Looms Under TUFS, Ministry of Textiles, Government of India
4. Textile Machinery Manufacture Association
5. Compendium of International Textile Statistics—2009–10
6. Fibre2Fashion News Desk—India

Knitting

15 Fundamentals of Knitting

Knitting is a method of fabric formation and is second only to weaving as a method of manufacturing textile fabrics, and in this method, yarns are interlooped to make fabrics.

The term *knitting* describes the technique of making textile structures by forming a continuous length of yarn into columns of vertically and horizontally intermeshed loops. Interlooping consists of forming yarn into loops, and these loops are held together by the yarn passing from one loop to the next loop. Knitting requires a relatively fine, smooth, strong yarn with good elastic recovery properties. The knitted structure gives good elastic properties to fabrics. Nowadays, Lycra is used to increase the elastic properties of knitted garments

The potentials of knitting technology are varied. The unique loop structure of knitting provides opportunities for using a minimum number of yarns.to make a fabric. By varying the size of loops, fabrics with different textures can be produced. Under tension, the shape of the loop will get distorted. Knitted fabrics can be engineered for extensibility and tailorability. Different types of yarns can be used to produce fabrics with different properties on each side.

15.1 COMPARISON OF KNITTING AND WEAVING

In knitting, fabric is produced by interlooping one set of yarns, whereas in weaving, fabric is produced by interlacing two sets of yarns. Even with a single yarn knitted fabric can be made. To make a woven fabric two sets of yarns are compulsory. Knitting fabrics are more stretchable than woven fabrics. Knitted fabrics are dimensionally unstable whereas woven fabrics are dimensionally stable. For knitting, yarns having medium strength are sufficient, whereas for weaving, yarns having good strength are necessary, especially for warp.

15.2 KNITTED FABRIC STRUCTURE

Knitted fabrics are built by loops formed by needles progressively row after row by intermeshing the loops. Knitted loops are arranged in rows and columns roughly equivalent to weft and warp of woven structures. Figure 15.1 shows a knitted fabric structure.

15.2.1 COURSE

The horizontal row of loops is called the course, and a course is formed by the adjacent needles during the same knitting cycle from the same yarn. For example in Figure 15.1, the continuous horizontal loops produced by yarn 1 together form course 1. Similarly, the horizontal loops produced by yarn 2 form course 2 and so

DOI: 10.1201/9780367853686-17

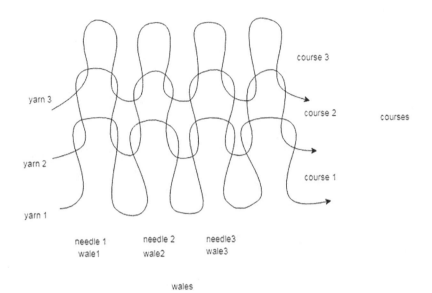

FIGURE 15.1 Knitted structure.

on. The length of yarn in one course is called course length. The course corresponds to weft in a woven fabric. The closer the courses, the denser the fabric and the higher the weight of the fabric per unit square area. Therefore, the number of courses per unit length will be a measure of the fabric texture or quality. Because the courses per unit length will change to a certain extent when the fabric is subjected to tension, it is not an accurate measurement of quality.

15.2.2 WALES

The vertical columns of loops are called wales, and a wale is the column of loops formed by the same needle in successive knitting cycles. For example in Figure 15.1, needle 1 forms wale 1, needle 2 forms wale 2 and so on. Each needle corresponds to one wale, and therefore, number of needles in a machine determines the number of wales in a fabric. The closer the needles are placed in a machine, the closer the wales will be and vice versa. The wales correspond to warp in a woven fabric.

15.2.3 STITCH DENSITY

Stitch density refers to the total number of loops in a unit area of the fabric, and it is expressed as the number of loops per square centimetre or square inch. Stitch density gives a measurement of fabric quality. Stitch density is obtained by multiplying the number of courses and number of wales per unit length:

Stitch density per square cm = Number of courses per cm × Number of wales per cm.

15.2.4   STITCH LENGTH

Stitch length, or loop length, is the length of yarn in one loop. The loop length in a knitted fabric will not change even if the fabric is subjected to distortion. It will be the same as when it is set in the machine. This is the only parameter which does not change in the knitted fabric during the subsequent processing. By adjusting the loop length, the weight per square area of the fabric can be changed and thereby the texture of the fabric. Hence, loop length is an important parameter in knitting. Figure 15.2 shows the stitch length of a loop.

15.2.5   MACHINE GAUGE

The number of needles per unit length in a knitting machine is called the gauge of the machine. This is a machine construction phenomenon, and it has to be decided during machine manufacturing itself. Machine gauge determines the wales per unit length. If the gauge of the machine is higher, that means more needles per unit length, and hence, the wales per unit length also will be higher. Fine yarns can be knitted in higher gauge machines to produce finer fabrics. Normally, in a 24-gauge machine (24 needles per inch), 30s to 40s yarn (English count) can be knitted. In a 20-gauge machine, 20s Ne yarn can be knitted conveniently.

15.3   TYPES OF KNITTING

There are two types of knitting:

1. Weft knitting
2. Warp knitting

Weft knitting is further classified into circular knitting and flat knitting. Again there are three types of circular knitting, namely plain, rib and interlock. Tricot, raschel, crochet and Milanese are the further classification of warp knitting. Figure 15.3 shows the detailed classifications of knitting.

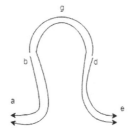

The length a b c d e is the loop length

FIGURE 15.2   Stitch length.

15.3.1 WEFT KNITTING

Weft knitting is the process in which the loops are formed progressively in the width-wise direction of the fabric. This is similar to inserting weft in a woven fabric and, hence, its name as weft knitting. In weft knitting, the needles form the loops sequentially in the same yarn one after another. Figure 15.4 shows the knitting progression in weft knitting.

15.3.2 WARP KNITTING

Warp knitting is the process in which the loops are formed progressively in the lengthwise direction of the fabric. This is similar to the movement of warp yarn in weaving and, hence, its name warp knitting. In warp knitting, all the needles form the loops at the same time. Figure 15.5 shows the knitting progression in warp knitting.

15.3.3 COMPARISON OF WEFT AND WARP KNITTING

1. In weft knitting, yarn feeding and loop formation occur at each needle in succession, whereas in warp knitting, yarn feeding and loop formation occur simultaneously in all the needles in a knitting cycle.
2. Weft knitted fabrics are mostly produced as tubular fabrics having a constant diameter. Warp knitted fabrics are knitted at constant width.
3. Weft knitting is more versatile and cost-effective, whereas warp knitting is limited to certain applications and more expensive.
4. Staple fibre-spun yarns and textured continuous filament yarns can be knitted in weft knitting machines, but knitting them in warp knitting machines is difficult.

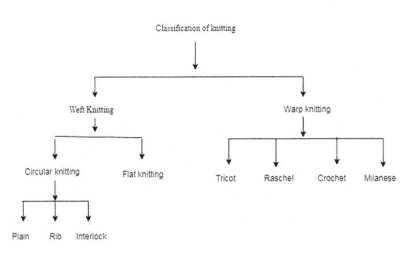

FIGURE 15.3 Classifications of knitting.

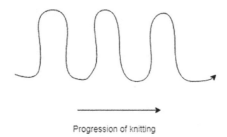

Progression of knitting

FIGURE 15.4   Knitting progression in weft knitting.

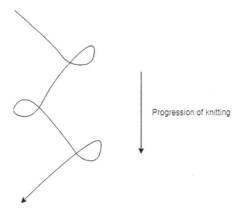

Progression of knitting

FIGURE 15.5   Knitting progression in warp knitting.

5. Even with a small quantity of yarn, weft knitted fabrics can be produced, but warp knitting requires a beam of yarn.
6. The yarn does not require sizing for weft knitting, whereas for warp knitting, the yarn has to be sized. However, waxing of the yarn is sometimes preferred for weft knitting.
7. Compared to weft knitted fabrics, warp knitted fabrics have more dimensional stability.
8. Both fine and coarse fabrics can be produced in weft knitting, whereas in warp knitting, generally fine fabrics only are produced.

16 Knitting Needles

Knitting needles are the important elements in loop formation in machine knitting. There are mainly three types of needles used in knitting:

1. Bearded needle
2. Latch needle
3. Compound needle

16.1 BEARDED NEEDLE

16.1.1 CONSTRUCTION OF BEARDED NEEDLE

The bearded needle is the first type of needle produced for machine knitting, and it is the simplest and cheapest needle. It is made from a single piece of metal strip. There are five main parts of a bearded needle and are shown in Figure 16.1

Stem 1 is the long, thin metal strip whose head 2 is bent downwards to form a hook or beard 3. The beard is used to pull the yarn into the loops. In the stem, a groove or eye 4 is cut to accommodate the tip of the beard during loop formation. The bottom of the stem is bent to form butt or shank 5, which will be placed in the groove of the cam during knitting operation.

16.1.2 KNITTING ACTION OF BEARDED NEEDLE

The knitting action of a bearded needle is illustrated in Figure 16.2. The needle has to be moved up and down for one knitting cycle. There are five important stages in a knitting cycle.

The needle is at rest position at A, with the previously formed loop 1 held by stem 5 in its hook portion. At B, the needle is raised, loop 1 is cleared from the hook and new yarn 2 is ready for presentation to the hook. At C, new yarn is fed to the hook, and presser 3 moves forward to close the hook. The needle starts moving downwards at D, and the new yarn is formed into a loop. The hook with the new loop starts moving through the old loop while the presser moves away from the hook. The needle moves further downwards at G, pulling new loop 4 through the old loop and at the same time the old loop is cast off or knocked over from the hook. After this, the needle is raised again for the next cycle of operation.

16.1.3 MERITS AND DEMERITS OF BEARDED NEEDLE

Bearded needles are the cheapest and simplest in construction. Very thin and fine bearded needles can be manufactured. Hence, fine fabrics can be produced using bearded needles. Because bearded needles require a presser for loop formation, it is

DOI: 10.1201/9780367853686-18

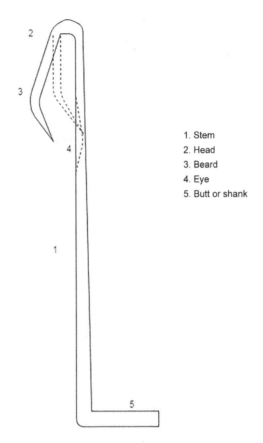

1. Stem
2. Head
3. Beard
4. Eye
5. Butt or shank

FIGURE 16.1   Main parts of bearded needle.

difficult to operate them individually. Due to this factor, bearded needles are used mainly in warp knitting.

16.2 LATCH NEEDLES

16.2.1 Construction of Latch Needles

A latch needle is a self-acting knitting needle. It is more intricate and expensive to manufacture. It has six main parts as illustrated in Figure 16.3.

The latch needle is made out of a thin metal strip whose one end is bend to form hook 1. Just below the hook portion, a slot is cut in the stem to accommodate latch 2. The latch is riveted in stem 4 by rivet 3, enabling the latch to have movement. Just above the other end of the stem, butt 5 has been attached. Below the butt, the stem extends as tail 6. The tip of the latch can be moved up so that it closes the hook.

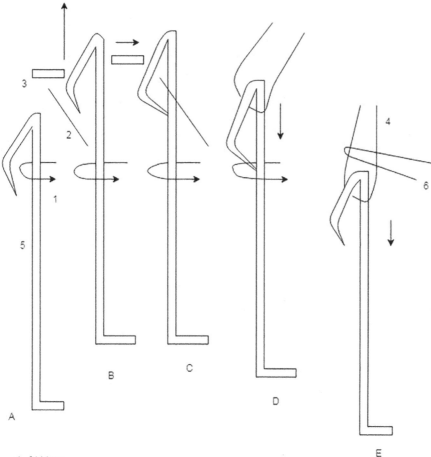

1. Old loop
2. New yarn
3. Presser
4. Newly formed loop
5. Needle stem
6. Old loop cast off

FIGURE 16.2 Knitting cycle of a bearded needle.

16.2.2 KNITTING ACTION OF LATCH NEEDLE

The knitting action or knitting cycle of a latch needle is illustrated in Figure 16.4. The needle requires an up-and-down motion to complete one knitting cycle. There are five positions of the needles during one knitting cycle:

    A. Rest position: Needle 1 is at the bottom position, and loop 2, formed in the previous knitting cycle, is held in the hook.

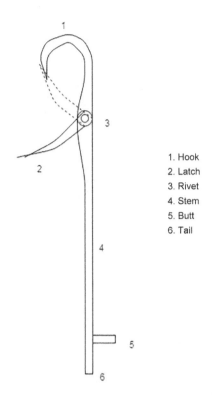

FIGURE 16.3   Latch needle.

B. Raising position: The needle is raised, and during this process, the old loop opens latch 3.

C. Feeding position: The needle is further raised, and new yarn 4 is fed to the needle hook. By this time, the old loop comes out of the hook completely.

D. Loop forming: The needle starts descending, and new loop 5 is formed by the yarn fed. The latch is closed by the old loop.

E. Knocking over: The needle descends farther, and the hook of the needle slips through the old loop, drawing the new loop into it. At the same time, the old loop slides off the hook, becoming part of the fabric. This is called knocking over the old loop. Now the needle is ready for raising for the next cycle of operation.

16.2.3   MERITS AND DEMERITS OF LATCH NEEDLE

The main advantage of latch needle is its self-acting nature, and it does not require any external element for its loop-forming action. The individual movement and control of the needle enable the stitch selection when knitting designed fabrics. The main disadvantage is its thickness due to the latch attachment. Hence, the production of fine fabrics is difficult. The latch needle is sturdy and used to produce courser fabrics without problem. It is mainly used in weft knitting.

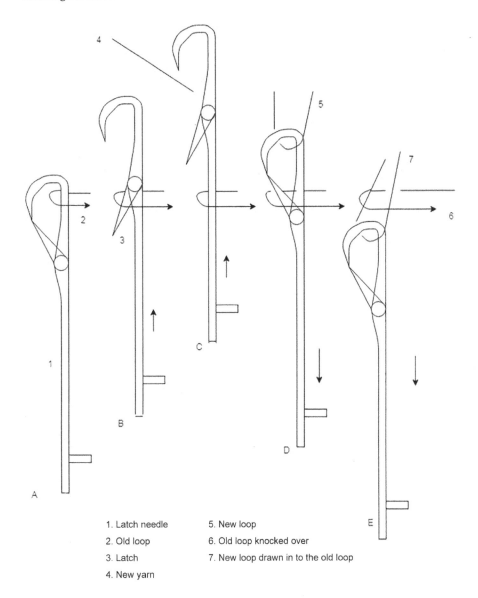

1. Latch needle   5. New loop
2. Old loop   6. Old loop knocked over
3. Latch   7. New loop drawn in to the old loop
4. New yarn

FIGURE 16.4 Knitting cycle of latch needle.

16.3 COMPOUND NEEDLES

16.3.1 CONSTRUCTION AND ACTION OF COMPOUND NEEDLES

Compound needles consist of two parts: open hook 1 and sliding closing element 2, as illustrated in Figure 16.5. The sliding element slides externally along a groove cut in the flat hook of the needle. The hook and the sliding element rise and fall together

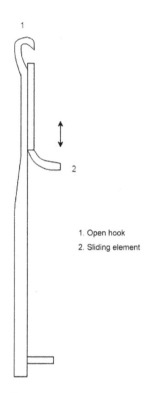

FIGURE 16.5   Compound needle.

during the knitting operation. But during raising, the hook raises faster to open the
hook. Once the yarn is fed into the hook, the hook again descends faster and closes
the hook. Then both the hook and the sliding element descend farther to make the
loop formation. Compound needles are mainly used in warp knitting because it is
easier to control the movement of sliding elements as one unit in a warp knit.

16.3.2   MERITS OF COMPOUND NEEDLES

Compound needles can be tapered to slimmer diameters of hooks, producing a larger
area inside the hook that is suitable to accommodate thicker yarn. Hence, compound
needles are used in fine, as well as course, gauge V-bed flat-knitting machines. Its
slim construction and short hook make it suitable for warp knitting at high speeds.

17 Principles of Weft Knitting

Weft knitting is the most diverse and widespread knitting technique that produces approximately 30% of the total apparel fabric production of the world. Weft knitted fabrics, ranging from inner garments to outer garments suitable for summer and winter, are mainly produced in circular weft knitting machines.

17.1 KNITTING ELEMENTS OF WEFT KNITTING

There are five important elements in a circular weft knitting machine:

1. Needle
2. Sinker
3. Cylinder or needle bed
4. Feeder
5. Cam

17.1.1 NEEDLE

The important function of loop formation and interlooping with adjacent loops are done by the needles in a knitting machine. The needles are the single most important element in the knitting machine without which no knitting can be done. The construction and workings of various knitting needles have been described in detail in Chapter 16.

17.1.2 SINKER

The sinker is the second important knitting element in a knitting machine. It is a thin metal plate cut to requirements as shown in Figure 17.1. It contains four important parts: nose, catch, butt and belly. Individual sinkers are kept between two needles, and separate cams give them forward and backward motion during knitting. They perform one or more of the following two important functions:

1. Holding down
2. Knocking over

The important function of the sinkers in modern single jersey knitting machines is to hold down the old loops in the needle stem when the needle raises. Holding-down sinkers are not necessary when two sets of needles are used as in the case of rib and interlock machines. In tricot warp knitting machines, sinkers help with knocking over of old loops.

DOI: 10.1201/9780367853686-19

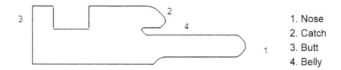

1. Nose
2. Catch
3. Butt
4. Belly

FIGURE 17.1 Sinker.

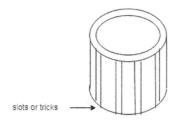

slots or tricks ⟶

FIGURE 17.2 Cylinder.

17.1.3 CYLINDER OR NEEDLE BED

A cylinder, or needle bed, is the main body of the knitting machine around whose surface the needles are placed. It is a metal cylinder, and on its surface, slots are cut to accommodate the needles as shown in Figure 17.2. The number of slots per unit length and its width depend on the gauge of the machine. The diameter of the cylinder determines the width of the fabric. The needles are placed in the slots or as it is called tricks of the cylinder. During knitting, the cylinder rotates along with the needles. One revolution of the cylinder corresponds to one course of knitting. In flat-knitting machines, instead of a cylinder, a flat rectangular metal bar is used.

17.1.4 FEEDER

Yarn feeders or guides are used in the knitting machine to place the yarn in front of the needle hooks so that during their movement, they catch the yarn by their hooks and form loops. The distance between the feeders and the needles are so important that too close a distance will break the needles often and too open will cause the needle hooks to miss the yarn. Therefore, it is very important to set the feeders firmly in the machine. Each feeder will guide one yarn, and hence, the number of feeders will determine the numbers yarns fed in the machine. To increase production, the number of feeders has to be increased in a machine. Figure 17.3 shows a feeder.

17.1.5 CAM

Cams are important mechanical elements which convert one type of motion into another type.

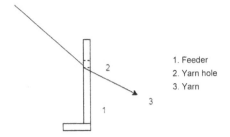

1. Feeder
2. Yarn hole
3. Yarn

FIGURE 17.3 Yarn feeder.

For moving the needle up and down cams are employed in knitting machines. The needle cam system consists of six cam segments:

1. Clearing cam
2. Stitch cam
3. Upthrow cam
4. Guard cam-1
5. Return cam
6. Guard cam-2

A knitting cam system employed in a single jersey knitting machine is illustrated in Figure 17.4. Knitting cams are attached either individually or in a unit form to a cam plate. For each feeder, a set of cams has to be arranged. The needle butts are placed in the groove of the cam. The needle enters from left to right in the cam as shown in Figure 17.4. Clearing cam 1 raises the needles to the required height so that the hooks are cleared from the old loops. Stitch cam 2 gives a downward movement to the needles so that loop formation takes place. Upthrow cam 3 brings the needle back to the rest position. Return cam 5 and guard cams 4 and 6 prevent the needle from falling out of the cam track.

The stitch length of the loop can be adjusted by adjusting the stitch cam. The stitch cam can be adjusted up and down in such a way that the downward movement of the needle can be adjusted. The more downward movement, longer the stitch length. The shorter the downward movement, shorter the loop length.

The sinker cam gives forward and backward movement to the sinkers. The sinker cam consists of three segments, namely race cam, withdrawing cam and sinker-return cam, as shown in Figure 17.5.

The sinker return cam can be adjusted according to the requirements.

17.2 KNITTING CYCLE OF SINGLE JERSEY WEFT KNITTING MACHINE

There are four important stages in the knitting cycle of single jersey weft knitting machines which are popularly known as plain knitting machines. These are illustrated in Figure 17.6.

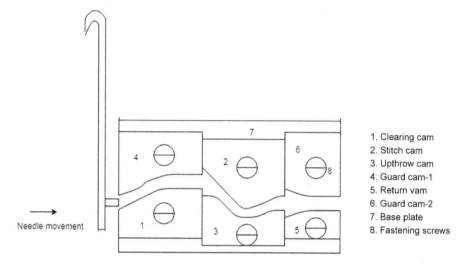

1. Clearing cam
2. Stitch cam
3. Upthrow cam
4. Guard cam-1
5. Return vam
6. Guard cam-2
7. Base plate
8. Fastening screws

Needle movement

FIGURE 17.4   Knitting cam.

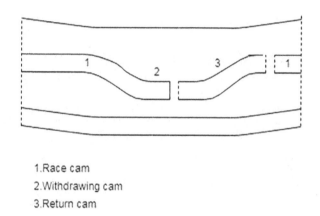

1. Race cam
2. Withdrawing cam
3. Return cam

FIGURE 17.5   Sinker cam.

A: Rest position: Sinker 1 is in the forward position, holding down the old loop, while needle 2 raises from the rest position.

B: Clearing: The needle has been raised to the maximum height, clearing old loop 3 from latch 4.

C: Yarn feeding: The sinker is partially withdrawn, and feeder 5 presents new yarn 6 to the descending needle, and the old loop slides on the needle stem ready to close the latch.

D: Knocking over: The needle descends to its bottom, drawing the yarn through the old loop to form new loop 7. At the same time, the old loop is knocked over from the hook and forms part of the fabric.

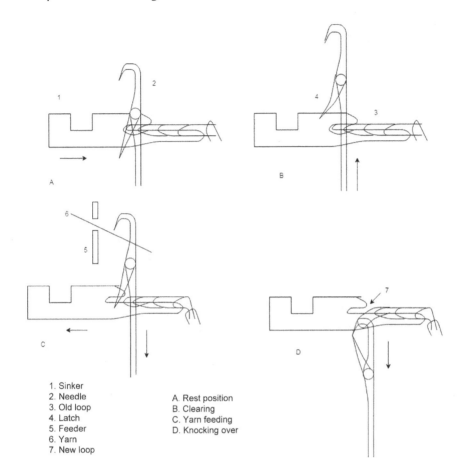

1. Sinker
2. Needle   A. Rest position
3. Old loop   B. Clearing
4. Latch   C. Yarn feeding
5. Feeder   D. Knocking over
6. Yarn
7. New loop

FIGURE 17.6   Knitting cycle of a single jersey circular weft knitting machine.

17.3   PLAIN CIRCULAR WEFT KNITTING MACHINES

Single jersey or plain knitted fabrics are produced in circular weft knitting machines. Because these fabrics have only one layer of loops in the cross section, they are called single jersey. Single jersey machines use latch needles. In these machines, the cylinder with needles revolves around stationary cams. The yarn is supplied from cones through the yarn feeder. The fabric in a tubular form is drawn from inside of the cylinder by tension and wound on fabric rollers. The winding mechanism also revolves with the cylinder. The essential parts of a circular plain knitting machine are illustrated in Figure 17.7.

Cylinder 2 is placed on cylinder bed 4 and is driven by pinion 8. Needles 1 is placed around the cylinder. Cams 3 are fixed on the cylinder bed around the cylinder, and they are stationary. Sinker bed 10 is fixed at the top of the cylinder, but it does not touch cylinder. Sinkers 5 are placed in the slot cut in the sinker bed. Feeders 6

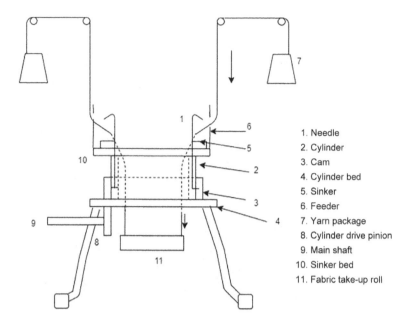

FIGURE 17.7 Schematic of a circular plain knitting machine.

are fixed around the sinker bed one for each cam unit. The yarn is drawn from supply package 7 and fed to the needles through the feeder.

When the cylinder revolves with the needles, due to action of the cam, the needles make up and down movements, and due to their knitting action, fabric is formed. The formed fabric is drawn by the fabric take-up roller 11 by tension and wound in the form of a roll. Figure 17.7a shows the knitting section of a circular plain knitting machine.

17.4 RIB KNITTING

Rib knitting is a technique of producing two layers of loops in the knitted fabric using two sets of needles. These fabrics are often called double jersey knitted fabrics due to the presence of two layers of loops. In the circular weft knitting machine, instead of sinkers, another set of needles are used, which are called dial needles. The dial needles are placed in the slots cut radialy in a circular dial perpendicular to cylinder needles. The fabrics produced from these machines have a vertical cord appearance on both sides of the fabrics, hence the name rib. The two sets of needles are placed alternatively or gated between each other, and this is called rib gating. The placement of the cylinder and the dial needles is shown in Figure 17.8.

The knitting action of a circular rib knitting machine consists of three stages and is shown in Figure 17.9.

A: Clearing: The cylinder and dial needles move forward to clear the cylinder needle and dial needle loops formed in the previous knitting cycle.

FIGURE 17.7(A)   Circular plain knitting machine.

B: Feeding: The cylinder needles and dial needles withdraw backwards, mak-
    ing the old loops close the latches. But at the same time, new yarn is fed to
    the cylinder and dial needles.
C: Knocking over: The needles are completely withdrawn enabling the old
    loops to cast off from the needle heads. At the same time, the cylinder
    needle head and dial needle head pull the yarn through the old loops, mak-
    ing new loops.

To give movement to the cylinder and the dial needles, separate cams for the cylinder and the dial needles are provided. There are two types of timings for the cams.
In synchronized timing, the cylinder and the dial needles pull the yarn at the same

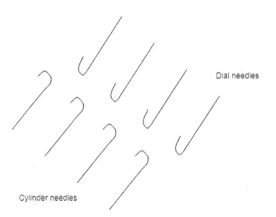

FIGURE 17.8   Rib gating.

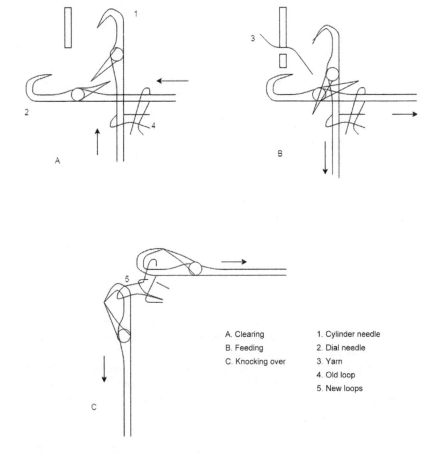

A. Clearing   1. Cylinder needle
B. Feeding   2. Dial needle
C. Knocking over   3. Yarn
                               4. Old loop
                               5. New loops

FIGURE 17.9   Knitting action of rib knitting machine.

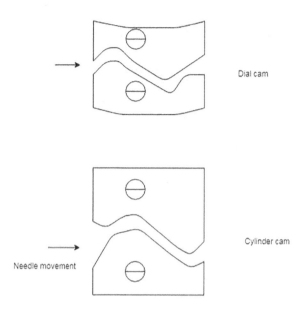

Dial cam

Cylinder cam

Needle movement

FIGURE 17.10 Synchronized timing cylinder and dial cams.

time to form loops. In delayed timing, the cylinder and the dial needles pull the yarn one after one another for loop formation. Figure 17.10 shows synchronized timing cylinder and dial cams.

17.5 INTERLOCK KNITTING

Interlock is a double jersey fabric having a technical face on both sides. It is a heavier and thicker fabric, having a relaxation of 30% in its width. To knit interlock fabrics, two sets of cylinder needles and two sets of dial needles are required. One set of cylinder needles will be longer, and the other set will be shorter. Similarly, one set of dial needles will be longer, and the other set will be shorter. The needles will be placed exactly opposite each other. The cylinder long needles will be placed opposite the dial short needles, and the cylinder short needles will be placed opposite the dial long needles. This is called interlock gating. Figure 17.11 shows interlock gating. Because of this gating, when the long cylinder needles knit, the short dial needles cannot knit, and when the short cylinder needles knit, the long dial needles cannot knit.

The cylinder and the dial cams are designed in such a way that the long needles will knit in one cam track and the short needles will knit in another cam track. Thus, for both the cylinders and the dials, two cam tracks each have been provided. Figure 17.12 shows an interlock cam system. Figure 17.13 shows an interlock cylinder cam.

The long needles of cylinder and dial knit at the first feeder yarn, and the short needles of cylinder and dial knit at the second feeder yarn. In Figure 17.12, when

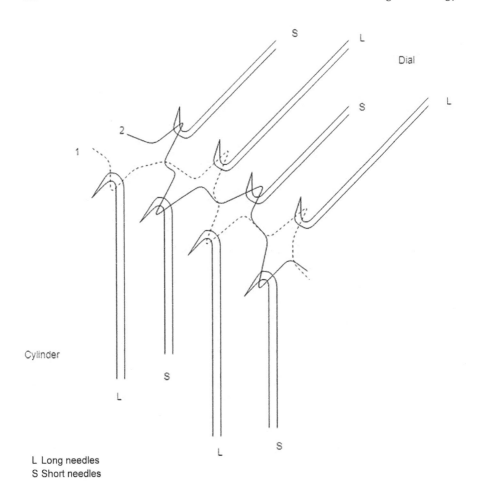

FIGURE 17.11 Interlock gating.

cylinder long needle 9 knits due to the action of cylinder cam 1 in feeder 1, the dial long needle 10 also knits after a delay due to the action of dial cam 6. When cylinder short needle 11 knits due to the action of cylinder cam 4 at feeder 2, the dial short needle 12 will also knit after a delay due to the action of dial cam 7. Interlock thus requires eight cams to produce one complete course.

17.6   PURL KNITTING

Purl structures have wales which contain both face and reverse loops. The simplest purl is 1×1 purl. Purl structures can be produced using double-ended latch needles

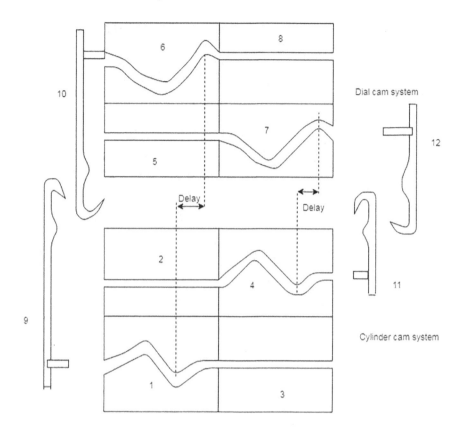

1. Cylinder long needle knit cam	5. Dial short needle idle cam	9. Cylinder long needle
2. Cylinder short needle idle cam	6. Dial long needle knit cam	10. Dial long needle
3. Cylinder long needele idle cam	7. Dial short needle idle cam	11. Cylinder short needle
4. Cylinder short needle knit cam	8. Dial long needle idle cam	12. Dial short needle

FIGURE 17.12   Interlock cylinder and dial cam system.

which have hooks at both ends. Two needle beds are required for purl knitting. The tricks of the two needle beds will be placed exactly opposite to each other in the same plane so that the double-ended needle can slide in both the tricks. Slides are used to move the double-ended needle from one bed to another. Figure 17.14 shows a purl knitting using sliders.

Knitting outwards from one needle bed the needle will produce a face loop with the newly fed yarn whilst the same needle knitting outwards with its other end hook from the other bed will produce a reverse loop. As the needle moves from one bed to the other bed, the old loop slides off the latch of the hook that produced it, and after sliding off, the old loop moves along the needle stem towards the other hook. But it

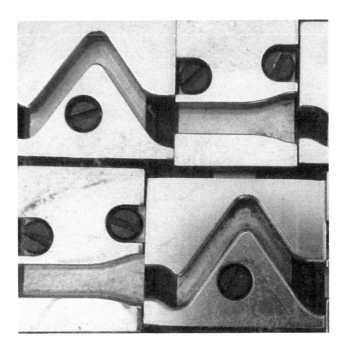

FIGURE 17.13 Interlock cylinder cam.

1. Front needle bed
2. Back needle bed
3,4. Sliders
5. Double-ended latch needle
6. Butt

FIGURE 17.14 Purl knitting using double ended latch needles.

cannot enter the other hook because it will pivot the latch to close. At the same time, a new yarn will be put in the hook, and when the needle moves backwards, the new yarn will make a new loop and so on.

18 Basic Structures of Weft Knitting

18.1 INTRODUCTION

There are four basic structures in weft knitting:

1. Plain
2. Rib
3. Interlock
4. Purl

To produce each structure different machines with different needle arrangements are required. Each primary structure may exist alone or in a modified way. Apart from the basic structures, many knitted structures are derived from these structures according to different requirements. Each structure has its own unique fabric properties and its own applications and end uses.

18.2 KNITTING NOTATION

Knitting notation is the simplest way of representing knitting structures. There are two types of notations. One is using point paper, and the other is using square paper. Point paper is used to create a running thread path notation for weft knitting structures, which is also recognized for warp knitting lapping diagrams. In point paper, each point represents a needle, and after drawing the thread path, it also represents the stitch. Each horizontal row of points represents adjacent needles during the same knitting cycle. Each vertical column of points represents the same needle.

In the square paper method, each square represents a needle or a stitch. An X symbol in a square represents a face stitch and O symbol in a square represents a reverse stitch. Figure 18.1 shows the point paper and square paper representations of face loop stitch and reverse loop stitch.

18.3 PLAIN STRUCTURE

The plain structure is the base structure, and its technical face is smooth, with the legs of the needle loops having the appearance of columns of V in the wales. On the technical backside, the heads of the needle loops will appear as semi-circles. Figure 18.2 shows a plain knit structure with notation. Plain knits can be unroved from the course knitted last. Plain knit structure is the simplest and most economical weft knit structure and has maximum cover factor.

DOI: 10.1201/9780367853686-20

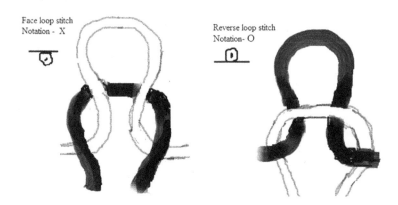

FIGURE 18.1   Face loop and reverse loop notation.

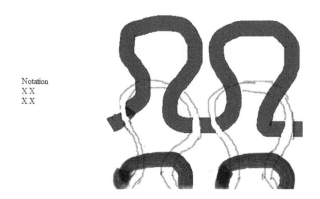

FIGURE 18.2   Plain knit structure.

18.4   RIB STRUCTURE

The simplest rib is 1×1 rib and it is knitted with two sets of latch needles. Rib is a double jersey fabric because it has two layers of loops. One layer of loops is formed by cylinder needles, and another layer of loops is formed by dial needles. It is a reversible fabric. Both sides have needle loop legs running vertically and give a rib appearance, hence the name rib knits. Rib fabrics have twice the thickness of plain knits and half the width. Rib is a balanced fabric, and it will not curl when cut. Figure 18.3 shows a rib structure with notation.

Rib structures are elastic and warm better than plain structures. Rib knits are best suitable for tops of socks, cuffs of sleeves and borders of garments.

18.5   INTERLOCK STRUCTURE

Interlock has the technical face of plain single jersey fabric on both sides, and due to this, both surfaces of interlock are smooth. Interlock cannot be stretched because the

FIGURE 18.3   Rib structure.

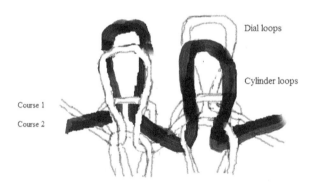

FIGURE 18.4   Interlock knit structure.

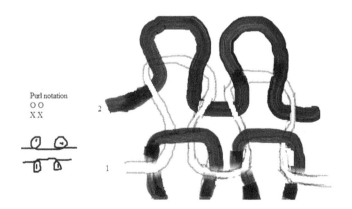

FIGURE 18.5   Purl knit structure.

wales on both sides are opposite to each other and locked, hence the name interlock. It is a balanced structure and lies flat without curl. Interlock is a thicker and heavier fabric. Figure 18.4 shows an interlock knit structure.

18.6 PURL STRUCTURE

Purl structure has a face loop stitch and a reverse loop stitch in the same wale. The simplest purl is 1×1 purl and consists of an alternate course of face and reverse loops. It has more lengthwise elasticity, and it is thicker than plain knit fabrics. Figure 18.5 shows a purl knit structure.

19 Types of Knitting Stitches

There are three types of stitches a knitting needle can produce during the knitting cycle depending on its movement:

1. Knit stitch
2. Float stitch
3. Tuck stitch

These stitches are used to create designs in fabrics.

19.1 KNIT STITCH

A knitted loop stitch is produced when a needle receives new yarn, makes a new loop and knocks over the old loop formed from the previous knitting cycle. This stitch is formed when the needle is raised to the height required for the latch to come out of the old loop and catch the new yarn so that it can make a new loop during its downward movement. The new loop is drawn through the old loop by the needle before casting off the old loop. This normal knitted stitch is called a knit stitch. This is the stitch which forms the bulk of knitted fabric.

19.2 FLOAT STITCH

When the needle is not raised to the required height for the needle to catch new yarn, the needle will not make new loop during its downwards movement. During the casting off of the old loop, the old loop of this needle gets straightened between the adjacent loops of both sides. This will look like a float. This stitch is called a float stitch or a miss stitch. A float stitch has the appearance of a U-shape on the reverse side of the fabric. Knitted fabrics with float stitches exhibit horizontal lines. More float stitches in a fabric will reduce widthwise elasticity and improve fabric stability.

Figure 19.1 shows a float stitch. Figure 19.2 shows the needle height for making a float stitch. A float stitch is represented as an empty square or as a bypassed point. A float stitch is also called a miss stitch.

19.3 TUCK STITCH

A tuck stitch is one in which the old loop is not cast off from the hook but receives the new loop. Thus, a tuck stitch will have two or more loops in the hook at the same time. Figure 19.3 shows a tuck stitch. When the needle is not raised to the required height for the latch to come out of the old loop, a tuck stitch is formed. Figure 19.2 shows the needle height for making a tuck stitch. A tuck is represented

DOI: 10.1201/9780367853686-21

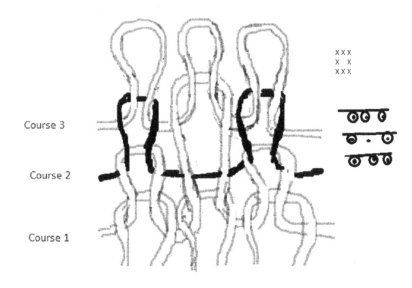

FIGURE 19.1   Float stitch.

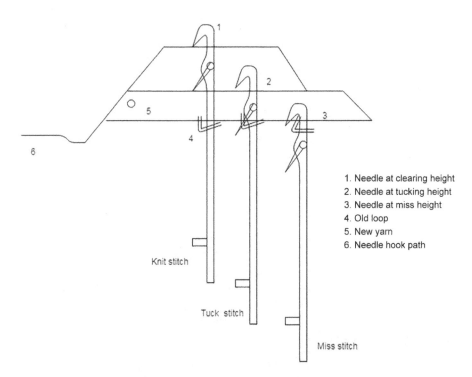

1. Needle at clearing height
2. Needle at tucking height
3. Needle at miss height
4. Old loop
5. New yarn
6. Needle hook path

FIGURE 19.2   Needle height for knit, float stitch and tuck stitches.

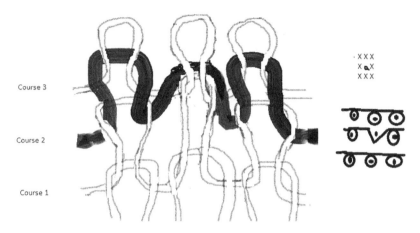

FIGURE 19.3   Tuck stitch.

as a dot in a square box or as a semi-circle over a point. The tuck stitch looks like an inverted V-shape.

A tuck stitch is formed by lifting the needle only halfway towards the clearing height. The old loop opens the latch but remains on the latch spoon and does not slide off on the needle stem. It remains in the needle hook, where a new loop joins with it before knocking over when the needle descends. Only a latch needle is capable of producing knit, tuck and miss stitches.

20 Patterning and Needle Selection

Patterns are produced in weft knitted structures either in the form of selected coloured yarns or by choosing different types of stitches. The width of the pattern in wales is determined by the number of needles selected, and the depth of pattern in courses is determined by the number of feeders with selection facilities. Horizontal strips, intarsia and plating are done by selecting yarns. Patterning by individual stitch selection is based on the principle that by altering the raising height of the latch needle, any one of the knit, miss and tuck stitches can be produced in the corresponding wale. These stitches will have different appearance and effect in the fabric.

There are many methods by which the needle height can be controlled:

1. Multitrack cams
2. Pattern wheel
3. Needle selection by punched tapes
4. Electronic needle selection

20.1 HORIZONTAL STRIPS

Horizontal striping is produced when coloured yarn is fed by selected feeder. The width of the strip depends on the number of feeders with coloured yarn. For example if feeders 1 to 4 are fed with, say, red colour and feeders 5 to 8 are fed with blue colours and the rest of the feeders fed with white colour, then a fabric with red colour stripe followed by blue colour stripe and then a white colour will be produced. Figure 20.1 shows a horizontal strip design. Alternatively, feeders may have the facilities to feed different coloured yarns one at a time. In these machines, each feeder will be provided with a number of yarn guides with different coloured yarns and a selection mechanism to select the required colour yarn. Figure 20.2 shows a four-colour selection facility for a feeder.

20.2 INTARSIA

Intarsia is a method of producing designs in pure colours in knitted fabrics. In intarsia, each block of needles is exclusively supplied with its own particular yarn to knit separate coloured areas. These yarns will run above the course and do not float behind the needle loops.

DOI: 10.1201/9780367853686-22

FIGURE 20.1 Horizontal strip design.

FIGURE 20.2 Four-colour selection facility.

20.3 PLATING

Plating is a technique in which two yarns of different characteristics like different fibres, colour or count are fed into the same feeder to make designs. Usually, the main yarn will continuously knit the ground fabric, but additional yarn will be fed along with the main yarn in selected feeders to produce design effect. Figure 20.3 shows a plating technique.

Yarn 2

Yarn 1

FIGURE 20.3   Plating technique.

20.4   MULTITRACK CAMS

In single jersey multi-cam track machines, needles with different butt heights will be employed. A four-track cam system will employ needles with four different butt heights. Every feed position will have fixed but exchangeable knitting, tucking and miss cams. A particular needle will be raised according to the cams positioned in that particular track. The total number of cam tracks influence the width of the repeat in the wales. Each butt position is used only once in the repeat so that the pattern width is equal to the number of butt positions. Figure 20.4 shows a four-track cam.

20.5   PATTERN WHEEL

A pattern wheel is a simple device on whose circumference tricks are cut. The gauge of the tricks will be equal to the gauge of the cylinder. In each trick pattern bits can be placed. The pattern wheel will be placed in each feeder at an angle to the cylinder, and they are free to rotate. When the needles move along with the cylinder, the needle butts will slip into the tricks of the pattern wheel. The pattern wheel will also rotate. As it rotates and since it is inclined, it will lift the needle butt according to the pattern bit placed in that trick. If there is no bit in the trick, the butt will be lifted to miss height as shown in Figure 20.5.

If there is a low bit, the butt will be lifted to tuck height. If there is a high bit, the butt will be lifted to clearing height. Thus, by placing the required pattern bit, the needle can be made to knit or tuck or miss. Figure 20.6 shows a pattern wheel mounted in a single jersey machine.

In Figure 20.5, a high bit has been placed at pattern wheel trick at 4 therefore the needle 1 has been lifted to clearing height so that a knit stitch is formed. A low bit has been placed at trick 5 so that needle 2 is lifted to tuck height. No bit has been placed at 6 so that the butt of needle 3 slips so that the needle does not raise to catch a new yarn and a miss stitch happens.

20.6   NEEDLE SELECTION BY PUNCHED TAPE

In punched-tape needle selection, a jack is used to lift the needle. Figure 20.7 shows a punched-tape needle selection system. Jack butt 2 is placed in the tricks of the

FIGURE 20.4   Multitrack cam.

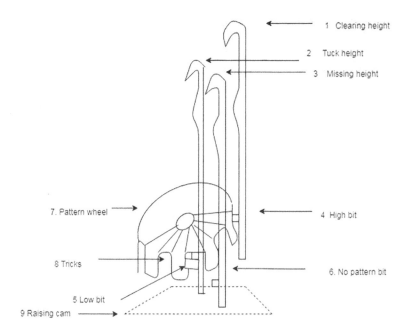

FIGURE 20.5   Schematic of a pattern wheel.

FIGURE 20.6   Pattern wheel mounted on a single jersey machine.

cylinder below needle 7. The jack butt is engaged with cam 3. When punched tape 6 with a hole is pressed against the tip of jack 1, as shown in Figure 20.7a, the tip of the jack will enter the hole; thereby, the jack will not have a lateral movement. As a result, the bottom portion of jack 4 will not engage with jack cam 5. Hence, the jack will not be lifted, the jack butt will not be moved out of the cam track, and the needle will follow the action of the cam. If there is no hole in the punched tape, as shown in

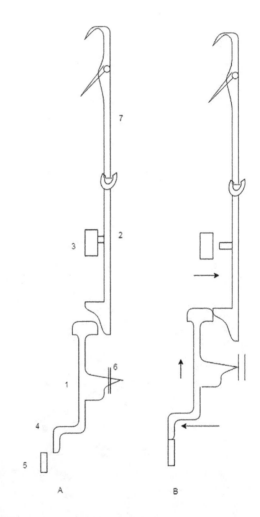

1. Jack
2. Jack butt
3. Cam
4. Lower portion of jack
5. Jack cam
6. Punched tape
7. Needle

FIGURE 20.7   Punched-tape selection system.

FIGURE 20.8   Electronic needle selection system.

Figure 20.7b, jack 1 will be moved laterally, the bottom portion of the jack will be engaged with cam 5 and the jack will be lifted. This will make the jack butt move outwards and disengage with cam 3. As a result the jack butt will not be lifted and hence the needle will also not be lifted.

20.7   ELECTRONIC NEEDLE SELECTION

Electromagnetic needle selection is used in latest machines. The electronic impulse that energizes an electromagnet is magnified, and this is used to select a needle through a mechanical means. These selection units are compact and can be fitted to a cylinder as well as to a dial needle selection. Figure 20.8 shows an electronic needle selection system.

21 Knitted Fabric Geometry

Weft knitted structures have the unique properties of formfitting and elastic recovery. Knitted loops have the ability to change shape when subjected to tension. This ability of the loops to change shape becomes a problem for maintaining dimensions of the knitted fabric during further processing, which causes size variation and shrinkage. By investigation HATRA (Hosiery and Allied Trades Research Association) found that the stitch length has an influence on knitted fabric dimensions.

21.1 STITCH LENGTH

Doyle P.J. (1953) discussed the relationship between stitch length and fabric dimensions. He showed that stitch density is primarily dependent on stitch length for a wide range of dry relaxed plain weft knitted fabrics and is independent of count, yarn structure and the system of knitting. This leads to the establishment of the following three basic laws relating to the behaviour of weft knitted fabrics:

1. Stitch length is the fundamental unit of weft knitted structures.
2. Loop shape determines the dimensions of knitted fabrics.
3. The relationship between loop length and loop shape can be expressed in the form of simple equations.

Munden D.L. (1959) derived the relationship between fabric properties and the yarn and knitting variables. Munden said that the plain knit structure essentially consists of the repeating unit of a single loop. The area of the fabric, its quality and its weight are all related and depend on the configuration and dimensions of the single loop. Munden showed that the knitted loop is a three-dimensional structure. In order to produce a flat knitted structure, the yarn is bent both in the plane of the fabric and in the plane right angles to the fabric. After knitting, the yarn, which was originally straight and desires to return to the straight state, is prevented from doing so by the equal and opposite reactions from the interlocking yarns. For any loop or row of loops to straighten, the greatest bending is necessary in neighbouring loops or row of loops in order to accommodate this change so that the whole structure tends to go into a state of minimum energy or minimum total bending. This is basically the mechanism of relaxation.

21.2 DIMENSIONS OF KNITTED FABRICS IN THE RELAXED STATE

The yarn is often temporarily distorted by the throw of the needle during knitting. Because the hosiery yarn is not perfectly elastic and the interlocking points are of

DOI: 10.1201/9780367853686-23

TABLE 21.1
Dimensional Constants for Plain Knitted Worsted Fabrics

Dimensional Constants	Dry Relaxed	Wet Relaxed
ks	19.0	21,6
kc	5.0	5.3
kw	3.8	4.1
R	1.3	1.3

high friction, the knitting strains are dissipated gradually on standing as the fabric tends towards its state of minimum energy.

The dimensions of the knitted fabrics in the relaxed state are determined by the knitting loop taking up its configuration corresponding to minimum energy. Munden suggested that this configuration is the geometrical property of the loop structure and is independent of the physical properties of the yarn or the amount of yarn knitted into the loop. Munden showed that fully relaxed plain knit fabrics constructed from yarns of different properties have similar loop shapes. This uniformity of loop shape is obtained only when the yarn has not been permanently deformed by the knitting action. Monofilament thermoplastic yarns, such as nylon, are invariably plastically deformed by the knitting action and in normal processing do not recover to their natural state. Their loop configuration will not, in general, resemble that of relaxed fabric.

Munden D.L. (1959) derived the following equations for thoroughly relaxed plain knitted worsted fabrics with his experimental results. In a relaxed condition the dimensions of plain knitted fabrics are given by the formulae:

1. Courses per inch (cpi) = kc/l,
2. Wales per inch (wpi) = kw/l,
3. Stitch density (S) = ks/l², and
4. cpi/wpi = R,

where l is the loop length in inch; kc, kw, and ks are dimensional constants and R is the loop shape factor. The k values for dry and wet relaxed plain knitted worsted fabrics are given in Table 21.1.

Using these equations, it is possible to predetermine the dimensions of relaxed knitted fabrics.

21.3   RELAXED STATES

Two formal relaxed states (Munden D.L. 1959) are recognized:

1. Dry relaxed state
2. Wet relaxed state

A dry relaxed fabric is one which is relaxed and has no wet treatment. A fabric is said to be in a dry relaxed state when all the stresses and strains introduced into the yarn during knitting have been dissipated with the exception of strains introduced by the yarn-to-yarn contacts required to maintain the shape of the loop. The usual procedure to achieve a dry relaxed state is to lay the fabric flat in the standard atmosphere and leave for the necessary time. A fabric is said to be wet relaxed when the fabric has been allowed in water until equilibrium is reached. This static soak is usually sufficient to fully relax the dry relaxed fabric. Fabrics knitted from hydrophilic yarns may be brought to a strain-free state by relaxation in water. Cotton fabrics which remain permanently distorted in the dry state recover completely from such strains when relaxed in water. All fabric measurements taken at wet relaxed state only are reliable.

21.4 TIGHTNESS FACTOR

Although the dimensions of relaxed knitted fabrics are dependent only on the loop length and are independent of the yarn count, the mechanical properties, fabric stiffness and pilling are dependent on the tightness of the construction. Munden D.L. (1962) pointed out that the ratio of natural yarn diameter to the loop length can be used to specify the tightness of the construction. Because the yarn diameter is proportional to linear density (tex), the tightness of plain knitted fabric is given by

fabric tightness = $K = \sqrt{T/l}$,

where T is yarn linear density in tex and
l is loop length in cm.

The fabric tightness is called the tightness factor, similar to the cover factor of woven fabrics.

The value of the tightness factor for plain knitted worsted fabric lies between 13 and 15.

21.5 LOOP SHAPE

Nutting T.S. and Leaf G.A.V. (1964) showed that the loop shape of weft knitted fabrics is controlled by loop length, fibre properties and method of relaxation. The loop shape is largely dependent on the ratio F/G, where F is yarn flexural rigidity and G is torsional rigidity.

21.6 POSITIVE FEEDERS

Because the fabric dimensions depend on the loop length, it becomes necessary to maintain a uniform loop length throughout the fabric in order to maintain the fabric quality. The uniformity of loop length depends on the uniformity of feed yarn tension. In modern knitting machines, due to the large number of feeders, the supply packages are kept separately, and the yarn is drawn through tubes to the feeders.

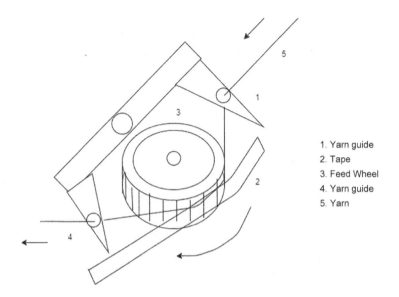

1. Yarn guide
2. Tape
3. Feed Wheel
4. Yarn guide
5. Yarn

FIGURE 21.1   Schematic of a positive feeder.

FIGURE 21.2   Positive feeder.

This causes uneven tension between feeders and uneven tension in the same feeder for a long period. Hence, positive feeders have been developed to draw the yarn from the supply packages positively and to feed at uniform tension in all the feeders. Figure 21.1 shows the schematic of a positive feeder. Figure 21.1 shows a positive feeder.

Continuous tape 2 encircles all feed wheel 3 and is driven by a single machine pulley. When the machine is running, the tape driven by the pulley drives all the feed wheels at a constant speed. Yarn 5, from the respective supply package, is drawn through guide 1 and taken around the feed wheel. When the feed wheel rotates, yarn is drawn at a uniform rate and fed into the feeder at a uniform tension. This uniform tension of yarn in all the feeders ensures a uniform loop length throughout the fabric during production.

REFERENCES

Doyle, P.J. (1953) "Fundamental aspects of the design of knitted fabrics" *Journal of the Textile Institute* Vol. 44. Pp. 561–578.

Munden, D.L. (1959) "The geometry and dimensional property of plain knit fabrics" *Journal of the Textile Institute* Vol. 50. Pp. T448–T471.

Munden, D.L. (1962) "Specification of construction of knitted fabrics" *Journal of the Textile Institute* Vol. 53. P. 628.

Nutting, T.S. and Leaf, G.A.V. (1964) "A generalized geometry of weft knitted fabrics" *Journal of the Textile Institute* Vol. 54. Pp. 45–53.

22 Flat Knitting

22.1 FLAT KNITTING MACHINES

Flat knitting is basically a weft knitting technique which uses flat needle beds. The popular V-bed flat knitting machines have two rib gated needle beds set at 90° between them giving an inverted V-shape appearance. The flat knitting machines follow an English system of gauging designated as E, which is the number of needles per inch. The flat knitting machine gauge ranges from E5 to E14. The knitting width of a flat knitting machine ranges from 14cm to 50cm for strapping and 80cm to 120cm for hand-operated garment machines. The width of automatic flat garment length machines ranges from 66cm to 240cm. The count range of yarns used will vary with the gauge of the machine. In a 12-gauge machine, 2/26s to 2/42s count may be used. In flat knitting machines, the needle bed is stationary, and the cam moves in a carriage in a slide along the full width of the machine. The trick walls of the needle beds are replaced with thin and polished knock-over bit edges. Latch-opening brushes are attached to the cam plates of both beds. These brushes open the latches of the needles. The yarn guide is attached to the carriage which slides along the full width of the machine in a rail.

Half-cardigan, full-cardigan and racked rib structures can be produced in flat knitting machines.

22.2 KNITTING ACTION OF FLAT KNITTING MACHINE

The knitting action of flat knitting machine consists of four stages. Figure 22.1 shows the four stages of the knitting action of flat knitting machines.

A. Rest position: At the rest position, the heads of the needles are level with knock-over bits. The butts of the needles are in straight line until contacting the raising cams.
B. Clearing: The needles are lifted to clearing height by the raising cams. The brushes open the latches of the needles.
C. Yarn feeding: At yarn feeding, a new yarn is fed to the needles as the needles start descending. Both needles pull the yarn to form a new loop.
D. Knocking over: As the needles descend farther, the knock-over of the front needle bed occurs after the knock-over of the back bed. A delayed timing of the knock-over is employed between front and back beds.

DOI: 10.1201/9780367853686-24

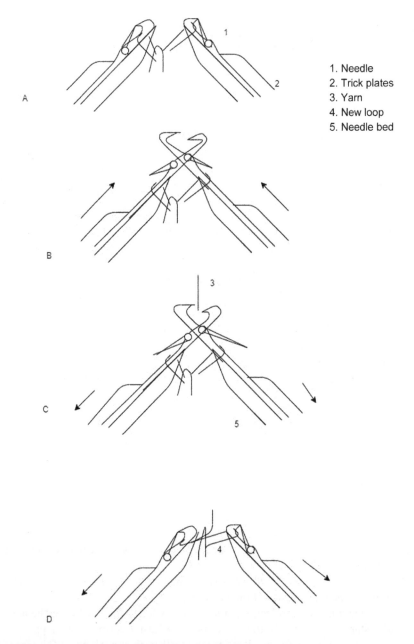

1. Needle
2. Trick plates
3. Yarn
4. New loop
5. Needle bed

FIGURE 22.1 Knitting action of a flat knitting machine.

23 Basics of Warp Knitting

In warp knitting, yarn feeding and loop formation will be done in all the needles simultaneously at every knitting cycle. Warp yarn passes from one needle loop in one course to another needle loop in the next course in the lengthwise direction.

23.1 OVERLAP

The needle loops are called laps in warp knitting. In order to produce a needle loop the yarn has to be swung upwards, shog it over the needle hook and swung downwards again. This movement of the yarn guide and the loop formed by it both are called overlap. The overlap is a shog across one needle hook only. In Figure 23.1, the movement abcd is an overlap.

23.2 UNDERLAP

The movement of yarn guide from one needle to another needle across the side is called underlap. It supplies yarn from one overlap to the next overlap. The underlap shog ranges from 0 to 3 needles spaces usually. Underlaps and overlaps are essential in warp knitting to join the wales of loops together. In Figure 23.1, the movement de is underlap.

23.3 CLOSED LAP

A closed lap is formed when a subsequent underlap shogs in the opposite direction to the preceding overlap. This will lap the same yarn around the front and back of the needle. Figure 23.2a shows a closed lap. Closed laps are more compact and heavier. They are more opaque and less extensible.

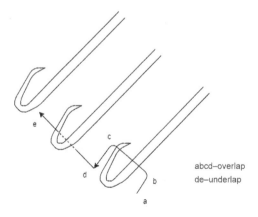

abcd–overlap
de–underlap

FIGURE 23.1 Overlap and underlap.

DOI: 10.1201/9780367853686-25

23.4   OPEN LAP

An open lap is formed when a subsequent underlap is in the same direction as the preceding overlap. An open lap will also be formed when an underlap is omitted. Figure 23.2b shows an open lap.

23.5   NEEDLE BAR

In warp knitting, all the needles have to move up and down simultaneously to form loops. For this purpose, all the needles are attached to a metallic bar, and this bar is called a needle bar. Similarly, all the sinkers will be attached to a bar for easy movement. The needle bar only will receive movements, and individual needles will not be lifted. Figure 23.3 illustrates a needle bar.

23.6   GUIDE BAR

The warp yarn guides, which are thin, metal plates drilled with a hole at the bottom for yarn passage, are held together by a metal lead at their upper end. The number of guides will be equal to the number of needles and are spaced as that of the needles. The leads are attached to a metal bar, which is called a guide bar. The guides will be hanging down from the guide bar, and each guide will be midway between two adjacent needles when at rest position. Each guide bar will receive warp yarns from

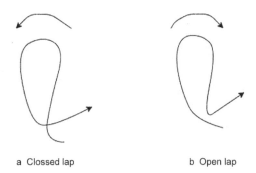

a Closed lap   b Open lap

FIGURE 23.2   Closed lap and open lap.

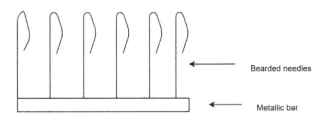

Bearded needles

Metallic bar

FIGURE 23.3   Needle bar.

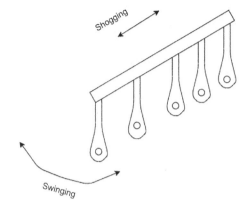

FIGURE 23.4   Guide bar.

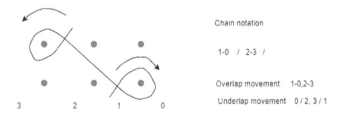

FIGURE 23.5   Lapping diagram and chain notation.

a warp beam. The minimum number of guide bars is two for normal warp knitting. Figure 23.4 illustrates a guide bar.

23.7 LAPPING DIAGRAM AND CHAIN NOTATION

The movement of the guide bars determines the overlaps and underlaps in warp knitted structures. The diagram showing the movement of guide bars is called a lapping diagram. The horizontal row of dots represents the needles, and each row of dots in a vertical direction represent successive courses. It is assumed that the pattern mechanism is on the right side. Because the guides always take a position between the needles, the spaces between the needles will be marked in numerical numbers starting from '0'. Because the underlaps always extend from the end of one overlap to the start of the next overlap, it is sufficient to draw the overlap only in the lapping diagram. In the chain notation, a dash (-) represents an overlap and a slash (/) represents an underlap. Figure 23.5 illustrates a lapping diagram and chain notation for a two-course repeat of a warp knit structure.

24 Warp Knitting Machines

24.1 TYPES OF WARP KNITTING MACHINES

There are four types of warp knitting machines:

1. Tricot warp knitting machines
2. Raschel warp knitting machines
3. Crochet warp knitting machines
4. Milanese warp knitting machines

Out of the four, only tricot and raschel machines are major types. Crochet machines use weft inlays apart from warp yarns. Milanese machines use two sets of threads in two guide bars to make open lap atlas traverses in the opposite direction without a return traverse up to the edges of the fabric. Due to their slow production, Milanese machines are rarely used.

24.2 TRICOT WARP KNITTING MACHINES

Tricot warp knitting machines mainly use bearded needles. Majority of tricot machines have only two guide bars, even though there are machines which have up to four guide bars. The sinkers are joined together, and they do the function of holding down, knocking over and supporting the fabric loops. Tricot machines produce mainly for lingerie and apparel with fine, closed knitted fabric. Tricot machines are usually fine gauge, raging from E28 to E44. Coarse-gauge machines are also available to knit stable fibre yarns. Figure 24.1 illustrates the knitting elements and warp beam arrangement in a tricot warp knitting machine.

The knitting cycle of tricot warp knitting machine is explained below.

There are seven stages in the knitting cycle of bearded needle tricot warp knitting machine. Figure 24.2 illustrates the knitting cycle of a bearded needle tricot warp knitting machine.

a. *Rest position:* The needle has risen from the knock-over position. The presser has withdrawn, and the guides are at the front of the machine. The sinker is in a forward position holding the overlaps.
b. *Backward swing and overlap shog:* The guides swing backwards and shog for overlap across the needles by one needle space.
c. and d. *Return swing and rise of needle*: The guides swing to the front, and at the same time, the needle rises to the full height, causing the newly formed overlaps to slip off the beards into the needle stem.
e. *Pressing:* The needles descends, and as soon as the new overlap comes inside the hook, the presser bar advances and closes the beard.

DOI: 10.1201/9780367853686-26

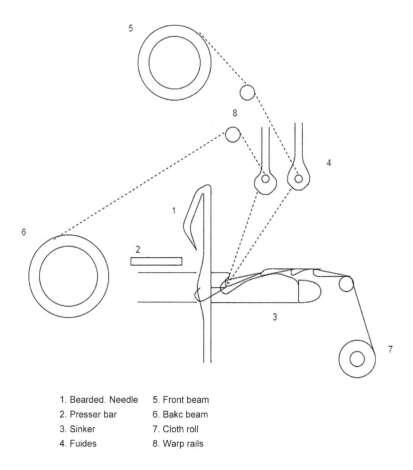

5

8

4

1

6

2

3

7

1. Bearded. Needle	5. Front beam
2. Presser bar	6. Bakc beam
3. Sinker	7. Cloth roll
4. Fuides	8. Warp rails

FIGURE 24.1 Knitting elements of tricot warp knitting machine.

    f. *Landing:* The sinkers move backwards, and the bellies land the old overlaps onto the closed beards. The presser is withdrawn.

    g. *Knock-over and underlap shog:* Continued descend of the needle makes the old overlap to cast off from the beards. The sinker moves forward to hold down the fabric loops and push them away from the needles. The guides shog for the underlap. The needles rise to the rest position.

24.3   RASCHEL WARP KNITTING MACHINES

Raschel warp knitting machine uses latch needles and have multi-guide bars. Dresses and household fabrics, trimmings, elastic tapes and curtains can be produced by raschel knitting machines. The fine-gauge raschel machines knit lightweight fabrics with a minimum elongation using 80dtex yarns, The gauge (needles per inches) of raschel machines vary from E1 to E32.

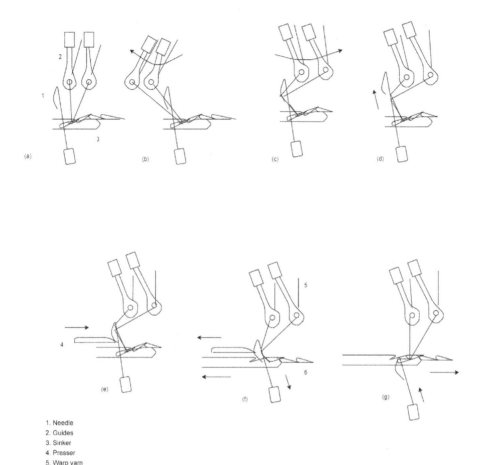

1. Needle
2. Guides
3. Sinker
4. Presser
5. Warp yarn
6. Fabric loops

FIGURE 24.2 Knitting cycle of bearded needle tricot warp knitting machine.

The latch needles are placed in a trick plate which extends to the full width of the machine. The warp beams are arranged above the machine. The sinkers are thin metal blades which move in horizontal plane above the trick plate. The fabric is drawn downwards and wound on the fabric roll. The sinkers are not joined together, and they perform the function of holding down the loops. Figure 24.3 illustrates the knitting elements of a raschel machine.

The knitting cycle of raschel warp knitting machine is explained below.

There are six stages in the knitting cycle of a latch needle raschel warp knitting machine. Figure 24.4 illustrates the knitting cycle of a latch needle raschel machine.

a. *Holding down:* Sinkers 3 move forward to hold the fabrics down while needles 1 start rising. Guides 4 are at the front of the machine.

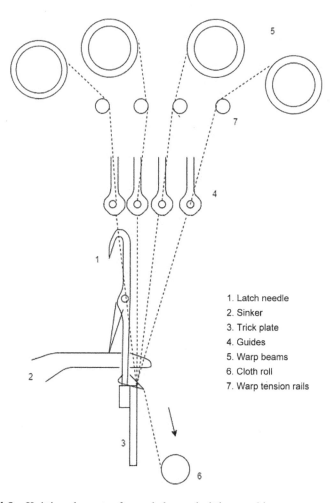

FIGURE 24.3   Knitting elements of a raschel warp knitting machine.

b. *Clearing:* The needles rise to their full height, and old overlaps slip down to the needle stem after opening the latches.

c. *Overlap:* The guides swing to the back of the machine and shog for the overlap. The sinkers start withdrawing.

d. *Return swing:* The guides swing forward while the warp threads wrap around the needle hooks.

e. *Latch closing:* The needles descend, and the old overlaps close the latches. The sinkers move forward.

f. *Knocking over:* As the needles descend farther, the needle heads pass below the trick plate, drawing the new overlaps through the old overlaps at the same time the old overlaps are cast off. The guides make the underlap shog, and the sinker moves over the trick plate to hold the fabric.

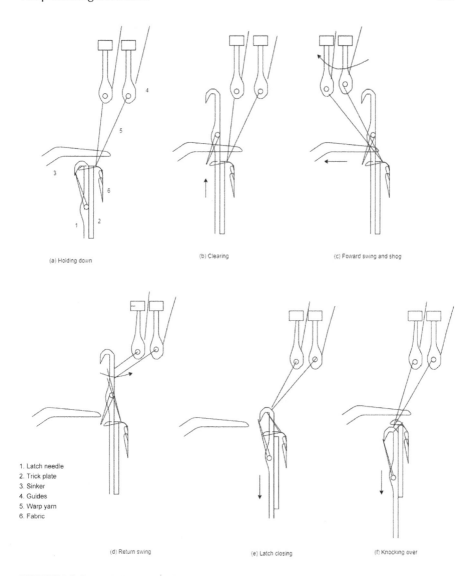

(a) Holding down

(b) Clearing

(c) Foward swing and shog

1. Latch needle
2. Trick plate
3. Sinker
4. Guides
5. Warp yarn
6. Fabric

(d) Return swing

(e) Latch closing

(f) Knocking over

FIGURE 24.4 Knitting cycle of a raschel warp knitting machine.

24.4 CROCHET WARP KNITTING MACHINES

Crochet machines use separate weft inlays to join the warp knitted wales to each other. A single horizontal needle bar with reciprocating action is used for forming wales. Instead of sinkers, a hold-back bar is used to prevent the fabric from moving out of the needles. Additional weft tubes are provided to supply weft inlay threads. The inlay bars are fitted above the needle bar and are shogged by pattern chains. Each needle is lapped by its own warp guide from below. The warp yarn may be supplied from individual yarn packages or by beams. Crochet machines are used to

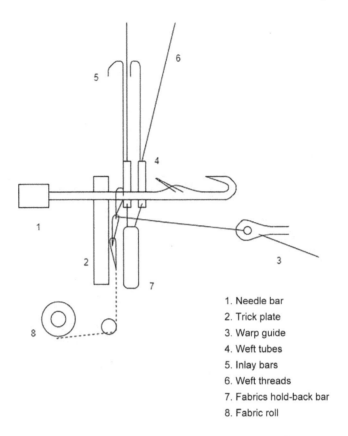

1. Needle bar
2. Trick plate
3. Warp guide
4. Weft tubes
5. Inlay bars
6. Weft threads
7. Fabrics hold-back bar
8. Fabric roll

FIGURE 24.5   Knitting elements of a crochet knitting machine.

produce fancy and open-work structures and narrow elastic laces. Crochet machines can use wide range of filament yarns from 20dtex to 1000dtex. Figure 24.5 illustrates the knitting elements of a crochet machine.

The knitting cycle of crochet machines have four stages. These are explained using the Figure 24.6.

   a. *Weft inlay:* The needle is in the withdrawn position. The weft inlay tubes are lowered, and weft is laid below the needle and above the warp thread.
   b. *Clearing the overlap:* Weft inlay tubes rise slightly on completion of their traverse, and the needles move forward to clear the old overlaps from its latches.
   c. *Overlap wrap:* The warp guides rise above the needles, make a shog and lower themselves to complete the overlap.
   d. *Knock over:* The needles move backwards and knock over the old overlaps at the same time drawing the new loop through the old overlap.

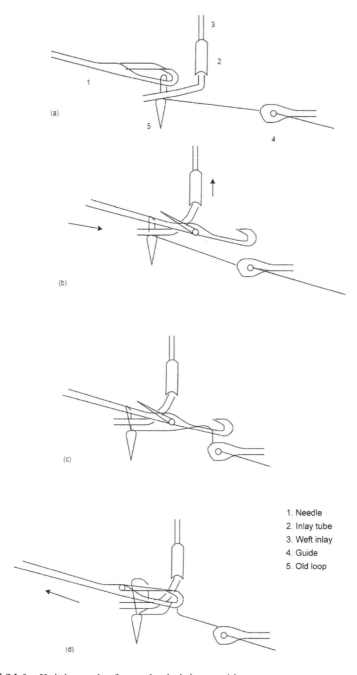

FIGURE 24.6 Knitting cycle of a crochet knitting machine.

25 Warp Knitted Structures

The majority of warp knitted structures are produced from two-guide-bar fine-gauge tricot machines. Multi-guide-bar raschel machines are used to produce laces, curtains and elastic fabrics successfully. Raschel machines can use up to 56 guide bars. Some of the important tricot structures are discussed in the following.

25.1 LOCKNIT

Of all warp knitted structures, locknit is the most produced structure and accounts for about 70% of total production. Its longer front guide bar underlap gives the fabric extensibility, cover, soft handle and good drape. Fine denier nylon filament yarn is used to produce finer fabrics in fine-gauge tricot machines. Elasticated fabrics for lingerie are produced from 40 denier nylon. Locknit fabrics can shrink up to 20%. The elasticity of locknit makes it suitable for intimate apparels and lingerie. Figure 25.1 gives the lapping diagram of a locknit structure.

25.2 REVERSE LOCKNIT

The reverse locknit has a shorter front guide bar underlap. Due to this, reverse locknit fabric has lesser extensibility and a shrinkage lesser than 10%. Its end uses are lesser than locknit fabric. Figure 25.2 shows lapping diagram of reverse locknit.

25.3 TWO-BAR TRICOT

Two-bar tricot fabric is the simplest warp knitted structure. The two guide bars cross diagonally between each wale. Two-bar tricot has poor cover. Figure 25.1 shows the lapping diagram of guide bars.

Front guide bar   Back guide bar

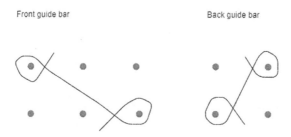

FIGURE 25.1   Lapping movement of a locknit.

DOI: 10.1201/9780367853686-27

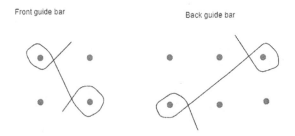

FIGURE 25.2   Lapping movement of reverse locknit.

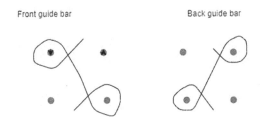

FIGURE 25.3   Guide bar lapping movement of a two-bar tricot.

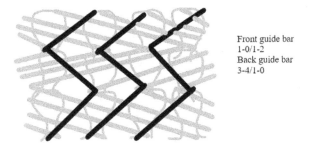

Front guide bar
1-0/1-2
Back guide bar
3-4/1-0

FIGURE 25.4   Sharkskin—technical back and chain notation.

25.4   SHARKSKIN

Sharkskin is produced by increasing the back guide bar movement to three or four needle spaces so that a long underlap is produced. This makes this structure more stable, rigid and heavier. Figure 25.4 shows the lapping movement and technical back of sharkskin fabric.

25.5   QUEENSCORD

In queenscord, the front guide bar makes the shortest underlap forming pillar stitches. The pillar stitches tie the longer back bar underlaps and allow lesser shrinkage of only 1% to 5%. The pillar stitch yarn gives a cord effect, hence the name queenscord. Figure 25.5 shows the technical back of queenscord fabric and chain notation.

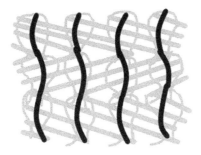

Front guide bar
1-0/0-1
Back guide bar
3-4/1-0

FIGURE 25.5   Queenscord—technical back and chain notation.

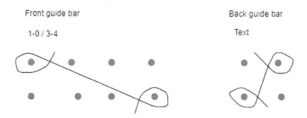

Front guide bar

1-0 / 3-4

Back guide bar

Text

FIGURE 25.6   Satin fabric—lapping diagram and chain notation.

25.6   SATIN

Satins are produced with larger front guide bar underlaps. This gives greater elasticity. Using filament yarn lustrous effect can be produced in satin fabrics. Figure 26.6 shows a lapping diagram of satin fabric.

Nonwoven

26 Nonwoven Batt Preparation Methods

Nonwovens are fabrics produced by arranging the fibres in a batt form and bonding them together. The American Society for Testing Materials defines *nonwoven* as "nonwoven is a textile structure produced by the bonding or interlocking of fibres or both, accomplished by mechanical, chemical, thermal, or solvent means or combination thereof. The term does not include paper or fabrics that are woven, knitted or tufted."

The main raw materials for nonwoven are polypropylene and polyester. Other fibres, such as viscose rayon, acrylic, polyamides and other specialty fibres, are used to a lesser extent. Nonwoven production involves two basic operations, namely batt preparation and bonding. Batt preparation involves arranging a thin layer of fibres in a web and then placing many layers of web one over the other in a form suitable for bonding. There are three categories of fibre-laying techniques for batt preparation:

1. Dry laid process: Parallel laying, cross laying and air laying belong to this category. Roughly about 50% of nonwoven production is based on dry laid processes.
2. Wet laid process: Water is used to lay the fibres.
3. Polymer laid process: Spun bonding and melt blown belong to this category.

Important batt preparation methods are discussed in the following sections.

26.1 PARALLEL LAYING

Cards are used to prepare the web. Card webs cannot be used directly for nonwoven production due to their low mass per unit area. Several card webs are placed one over the other to the required thickness to form a batt. The cards are placed at right angles to the lattice. The card webs are turned 90° using a guide plate and placed one over the other in the long lattice as shown in Figure 26.1. The number of cards depends on the thickness of the batt and the web areal density.

In parallel laying, all the card webs are parallel to each other, and most of the fibres lie parallel to the batt. Very few of them lie across the batt. Hence, parallel laid nonwovens have more tensile strength in the length direction or machine direction, and they are weak in the cross direction. This has reduced the use of parallel laid nonwovens to applications in which strength is necessary in one direction only. Another disadvantage of the parallel laid nonwovens is their width restriction. The width of the fabric cannot be more than the width of the cards.

DOI: 10.1201/9780367853686-29

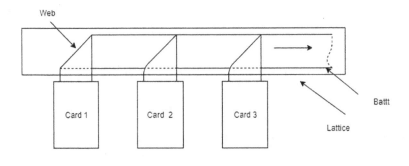

Parallel Laying

FIGURE 26.1   Parallel laying.

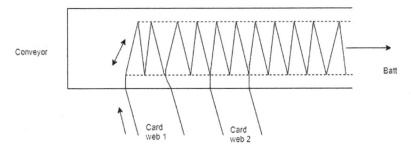

FIGURE 26.2   Cross laying.

26.2   CROSS LAYING

The cards are placed at right angles to the lattice. The card web is traversed back and forth across the lattice width while the lattice is moving forward. As a result, the web will be laid in a zigzag manner in the lattice. If the lattice moves faster the batt thickness will be less and a slower lattice movement will result in more thickness of batt. Figure 26.2 shows a cross laid batt.

There is a problem in cross laying: the edges of the batt will be laid heavier than the middle. The majority of fibres in cross laying lie in the cross direction. Consequently, cross laid fabrics are strong in the cross direction and weak in the machine direction.

26.3   AIR LAYING

In the air laying method, air is used as a medium for laying the fibres in the batt. Figure 26.3 shows an air laying process. The fibres are opened in opening machines and are fed to hopper 1. Opening roll 2, revolving at high speed, opens fibres 3 further and delivers at the back. Strong airstream 4 takes the fibres from the surface of the opening roll and deposits them on perforated conveyor 6, forming batt 7. The remaining air goes into suction 5.

In air laying, the fibres fall on the batt in an inclined plane that is at an angle to the plane of the fabric. Due to this, air laid fabrics have better recovery from compression than cross laid fabrics.

26.4 WET LAYING

Water is used as a medium for laying fibres. The fibres are cut very short (6–20mm) and dispersed in a large quantity of water. The amount of water should be enough to prevent fibre aggregating. Figure 26.4 illustrates a wet laying system. Fibre water dispersion 1 is passed in a trough. Take-up roll 2 takes the fibres and deposits them on conveyor 3. Extra water is drained through water drain 4. The batt is dried by heaters 5.

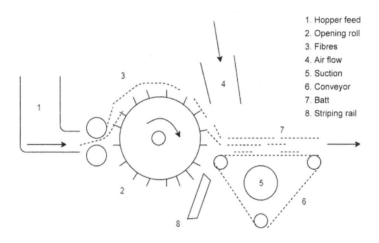

1. Hopper feed
2. Opening roll
3. Fibres
4. Air flow
5. Suction
6. Conveyor
7. Batt
8. Striping rail

FIGURE 26.3 Air laying.

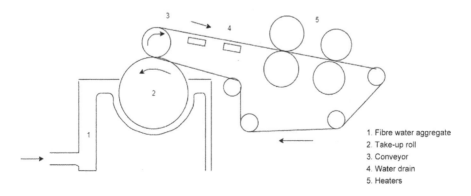

1. Fibre water aggregate
2. Take-up roll
3. Conveyor
4. Water drain
5. Heaters

FIGURE 26.4 Wet laying.

Wet laying is used to produce disposable products, such as hospital drapes and gowns and single-use filters.

26.5   SPUN LAYING

Spun laying process involves extrusion of the fibres from the raw material, drawing them and laying them into a batt. Often laying and bonding are done as a continuous process. Polyester and polypropylene are the main raw materials. Developments in spun laid process make it more versatile for producing soft-feel lightweight fabrics. Figure 26.5 shows the spun lain process.

The polymer chips are fed to the hopper and are melted. Polymer melt 1 is fed to extruder 2, which, in turn, feeds the polymer melt to spinneret 3 with pressure. The polymer melt coming out of the minute holes of the spinneret is solidified into fibres by cooling air, 4. Then the fibres are drawn by a high-velocity airstream, 5, through a tube. The fibres drawn by the airstream are deposited on moving conveyor belt 7. Due to the movement of the conveyor, the fibres will have a strong machine direction orientation. By oscillating the air tubes forwards and backwards, a cross-direction orientation to the fibres can be given. The thickness of the batt can be controlled by adjusting the speed of the conveyor and the number of fibres drawn from the spinneret.

26.6   MELT BLOWN

Melt blowing is the process of producing very fine fibres without the use of fine spinnerets. The melted polymer is extruded through relatively larger holes. When the

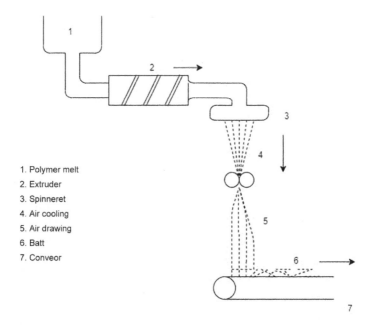

1. Polymer melt
2. Extruder
3. Spinneret
4. Air cooling
5. Air drawing
6. Batt
7. Conveor

FIGURE 26.5   Spun laid process.

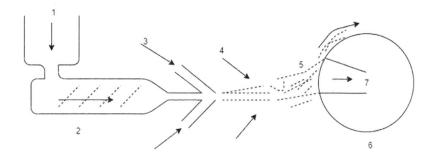

1. Hopper
2. Extruder
3. Hot air
4. Cold air
5. Batt
6. Collector
7. Suction

FIGURE 26.6 Melt blown process.

polymer melt leaves the extrusion holes, a high-speed stream of hot air is blown over it which breaks up the flow and stretches the fibres until the fibres are very fine. Then cold air is blown over it, and the polymer solidifies. During this process at some point, the filament breaks into stable fibres. The fine, stable fibres are collected in a conveyor as a batt. Figure 26.6 shows a melt blown process.

In melt blown process, the fibres are extremely fine, and fibre-to-fibre contact is more resulting in a batt with greater integrity. For many end uses, such as ultrafine filters for air conditioners, personal face masks, hygiene products and oil spill absorbents, the batt as such can be used.

27 Nonwoven Bonding Techniques

The second-most important process in nonwoven production is bonding the fibre. The property of the nonwoven product depends on method used for bonding to a larger extent. Different bonding methods produce different effects on the final product. Bonding is carried out as a separate process. But in spun laying and melt blown, bonding is carried out as a continuous process. The major fibre bonding techniques are as follows:

1. Mechanical bonding: Mechanical bonding rely on frictional forces between fibres and fibre entanglement. Needle punching, stitch bonding and hydro-entanglement belong to this category.
2. Thermal bonding
3. Chemical bonding

27.1 NEEDLE PUNCHING

The concept of needle punching is simple. Needles with barbs cut in their sides are penetrated into the batt. During penetration, the barbs catch some fibres and bend them in the cross-sectional direction. The bent fibres will be left as such by the needles during the return movement. This creates a bonding between fibres. Figure 27.1 illustrates a needle punching machine.

Large numbers of needles are pegged into needle board 1. The needle density will vary according to the fibre and the product. The needle board is reciprocated up and down by crank mechanism 2. Batt 7 is fed by feed roll 5 into the gap between stripper plate 3 and bed plate 4. The batt will be punched by the needles when the needle board descends and the fibres caught by the barbs will be pulled through other fibres. When the needle board returns, the loops formed by the downstroke will be left as such in the form of loops. The punching converts the batt into a felt. The felt will be delivered by delivery roll 6. A typical needle punching machine will have a stroke frequency of 120 strokes per minute, a depth of penetration of 10mm, a thickness between two plates of 25mm and a needle density of 8000 needles per square metre.

In needle punching, only vertical loops or pegs are formed. This alone will not be sufficient to give strength to the fabric. Parallel laid batts are not suitable for needle punching. Cross laid, air laid and spun laid batts are mostly used in needle punching.

Needle felts have good breaking tenacity and high tear strength. Their modulus is low and recovery from extension is poor. Needle felts are used in gas filtration and wet filtration. Due to their homogeneity nonwoven filters are better than woven

DOI: 10.1201/9780367853686-30

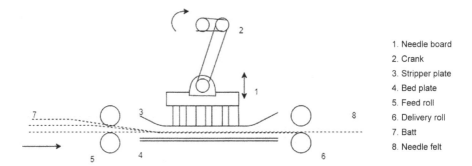

FIGURE 27.1   Needle punching machine.

fabric filters. The yarns in the woven fabric tend to stop the flow, and only the spaces between the yarn allow filtration. Needle felts are used as carpets widely.

27.2   STITCH BONDING

Stitch bonding is a technique in which the nonwoven batts are strengthened by knitted stitches in a warp knitting machine. Mainly cross laid and air laid batt are used in stitch bonding. A modified compound needle warp knitting machine is used. Figure 27.2 illustrates a stitch bonding technique.

Batt 5 is passed between specially designed needles 1 and guides 3. The needles, during their forward movement, penetrate the batt, and the hooks are open. Guides 2 will lap the stitching threads 4 in the hook. Then the needles are withdrawn, the hooks get closed by the tongue, the old loops are knocked over and new loops are formed. Two stitches, namely pillar stitch or tricot stitch, are used in stitch bonding. Stitch bonded fabrics are soft and flexible. These are used as backing fabrics for lamination and covering material in mattresses.

Sometimes stitch bonding is done without threads. The needles move backwards and forwards without lapping the threads. During backwards movement, the needles pick up some fibres from the batt and form loops. During the second cycle, the newly formed loops are pulled through the previous loops just in normal knitting. These fabrics are used in insulation and in decoration.

27.3   HYDRO-ENTANGLEMENT (SPUNLACE PROCESS)

The process of hydro-entanglement involves hitting a high-pressure jet of water on the web which produces an entanglement of the fibres. Figure 27.3 illustrates a hydro-entanglement process.

Batt 3, to be bonded, is passed over perforated drum 4. The high-pressure water jets 2, emanating from nozzle 1, hit the batt, and the fibre ends are twisted to entanglement. The water stream should be just like needles to produce the desired entanglement in the batt. The water droplets, after hitting the batt, are collected by the vacuum through the perforated drum and recycled.

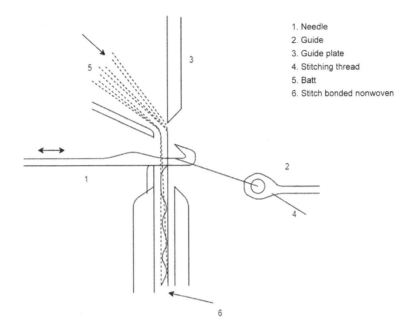

1. Needle
2. Guide
3. Guide plate
4. Stitching thread
5. Batt
6. Stitch bonded nonwoven

FIGURE 27.2   Stitch bonding.

The batts which are too light for processing in needle loom are successfully bonded through the hydro-entanglement process. Hydro-entangled products are used as wipes, surgeons' gowns and disposable protective clothing. Hydro-entangled products are lint-free because the lint is carried away by the water droplets.

27.4   THERMAL BONDING

In thermal bonding, heat is applied up to the melting point of the thermoplastic fibres so that the fibres melt and stick together. Three types of fibrous raw materials can be used in thermal bonding. First, all the fibres can be of the same melting point, and heat may be applied at localized points so that bonding is created at those points. Second, a blend of low-melt fibre with either a fibre having higher melting point or a non-thermoplastic fibre can be used. If polypropylene is used as a bonding fibre, the temperature to be raised is 125 °C to 155 °C, and for polyethylene the temperature is 90°C to 110 °C. Third, a bicomponent fibre, having a core with a higher melting point and a sheath with a low melting point, can be used. All methods of batt production are suitable for thermal bonding except the wet laid process. However, spun laying with point bonding is an ideal process. Figure 27.4 shows a thermal bonding process.

Batt 2, suitable for thermal bonding, is passed through heated calendar rollers 1. The calendar roller temperature should be enough to melt the bonding fibres. Pressure will be applied between the calendar rollers according to the requirement. After bonding, the thermal-bonded nonwoven will be delivered.

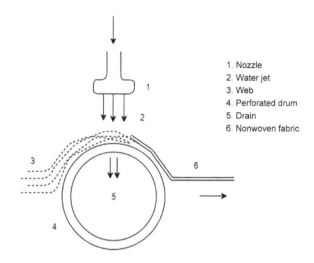

1. Nozzle
2. Water jet
3. Web
4. Perforated drum
5. Drain
6. Nonwoven fabric

FIGURE 27.3 Hydro-entanglement process.

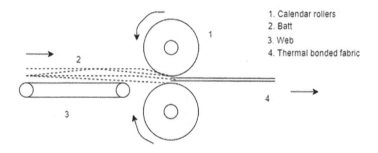

1. Calendar rollers
2. Batt
3. Web
4. Thermal bonded fabric

FIGURE 27.4 Thermal bonding process.

27.4.1 THROUGH-AIR BONDING

The batt is passed through a hot-air oven so that the fusible fibre will melt and create a bond. This technique is used to produce high-loft textiles.

27.4.2 THERMAL BONDING WITH PRESSURE

The batt is passed through the hot-air oven and is calendared by two heavy rollers to bring to the required thickness.

27.4.3 THERMAL BONDING WITH HIGH PRESSURE

The batt is passed through two heated calendar rollers. Pressure is applied between the rollers so that the bonded fabric is dense and heavily bonded. These nonwovens will have good strength, very high modulus and stiffness with good recovery from bending. These nonwovens are used as geotextiles, stiffeners and in filtration.

27.4.4 THERMAL BONDING WITH POINT CONTACT

Fabrics produced with bonding all over the batt is too stiff. To reduce the stiffness, point bonding is preferred. To produce point bonding, one of the calendar rollers is engraved with a pattern so that the effective contact between the two rollers is reduced. The bonding is confined only to the contact points. Fabrics produced by point bonding is flexible and relatively soft. The fabrics are used as geotextiles, filtration medium and protective clothing.

27.5 CHEMICAL BONDING

In chemical bonding the batt is treated with a bonding agent, dried and cured at a higher temperature. The curing temperature is 120 °C to 140 °C for 2 to 4 minutes. The curing develops cross-linking in the bonding agent and creates bonding. Synthetic lattices, such as acrylic latex, styrene-butadiene latex and vinyl acetate latex, are used as bonding agent. There are many methods of binder application. Each method produces different properties. They are discussed in the following sections.

27.5.1 SATURATION BONDING

In saturation bonding, the whole batt is wet with bonding agent, and all fibres are covered with a film of binder. Figure 27.5 shows saturation bonding process. Batt 5 is passed under an impregnation roller which is immersed in binder solution. As the batt moves, it picks up binder, and the excess binder is squeezed by the squeezing rollers. After squeezing, the impregnated batt is thoroughly dried in the drying chamber. In through drying, hot airstream is blown down through the fabric. During drying, all the water evaporates, leaving the binder particles on the fibre surface in a film form. The dried batt is cured in a separate compartment with the required temperature. Saturation bonded fabrics are stiff and have less tensile strength. They are used as interlining fabrics and filters.

27.5.2 FOAM BONDING

Because saturation bonding involves a lot of water, foam bonding has been developed. The binder solution and air are passed through a turbine which beats the two into a foam. The foam is applied to the batt by the impregnating roller as shown in Figure 27.6. After impregnation, the batt is dried and cured.

27.5.3 PRINT BONDING

In print bonding, the binder is printed with a printing roller or a rotary screen printer. Only the printed areas of the batt will be bonded, leaving the non-printed areas free. Print-bonded fabrics are softer and flexible, owing to the large unbonded areas.

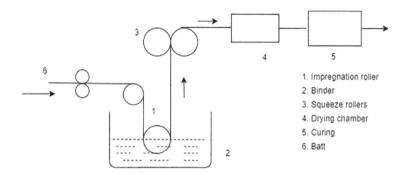

FIGURE 27.5   Saturation bonding process.

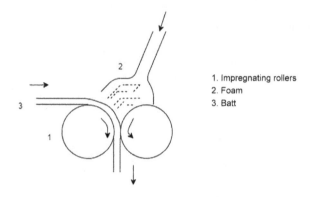

FIGURE 27.6   Foam impregnation.

27.5.4   Spray Bonding

In spray bonding, the latex is applied by spraying using spray guns. The spray penetrates about 5mm into the surface. The batt is revered and sprayed with the binder. After drying and curing, a thick, open and loft fabric is produced.

Index

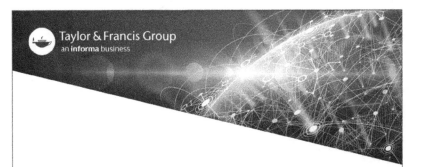

—